Newnes Guide to Radio and Technology

Newnes Guide to Radio and Communications Technology

Ian Poole

Newnes

AMSTERDAM   BOSTON   HEIDELBERG   LONDON   NEW YORK   OXFORD
PARIS   SAN DIEGO   SAN FRANCISCO   SINGAPORE   SYDNEY   TOKYO

Newnes
An imprint of Elsevier
Linacre House, Jordan Hill, Oxford OX2 8DP
200 Wheeler Road, Burlington, MA 01803

First published 2003

British Library Cataloguing in Publication Data
A catalogue record for this book is available from the British Library

ISBN 0 7506 56123

CDMA2000® is a registered trademark of the Telecommunications Industry
Association (TIA-USA)
•cdmaOne™ is a registered trademark of the CDMA Development Group
Bluetooth™ is a trademark owned by Bluetooth SIG, Inc.

For information on all Newnes publications visit our website at
www.newnespress.com

Composition by Genesis Typesetting Limited, Rochester, Kent

Transferred to Digital Printing 2010

Contents

Preface   vii

1 An introduction to radio   1

2 Radio waves and propagation   17

3 Modulation   51

4 Antenna systems   76

5 Receivers   119

6 Transmitters   179

7 Broadcasting   204

8 Satellites   221

9 Private mobile radio   238

10 Cellular telecommunications   247

11 Short-range wireless data communications   282

Index   291

Preface

Radio technology is a fascinating subject that encompasses an enormous number of topics. In recent years its importance has risen tremendously. Originally its uses were restricted and normally licences were required. Now it is regulated less and with the widespread use of cellular telephones, wireless local area networks and much more, it is used in a far wider range of applications.

The aim of this book is first to provide a grounding in the principles that underpin radio, and then to provide understandable introductions to various radio applications from broadcasting to satellites and cellular telecommunications to wireless local area networks.

Many of the basic principles have been long established, but the new technologies are advancing at a tremendous rate. Possibly the fastest changing area is that of cellular telecommunications. The first main systems started to appear in the 1980s. During the 1980s their use was restricted mainly to businesses as the costs of owning and using these phones was high. However, this has all changed and they are widely used across sectors of business and society. It is the aim of this book to look at today's latest technologies so that the reader is able to gain a good appreciation of what is actually being used now.

In preparing this book it has been necessary to seek the advice and assistance of several other people. I am indebted to many that have helped by discussing technologies, providing reference material, reviewing the work and supplying images. The support, assistance and encouragement of all has been very much appreciated, and has been invaluable in the preparation of the book.

Ian Poole
January 2003

1 An introduction to radio

Radio technology is an integral part of our everyday lives. Since it was first invented, its importance has steadily increased and today we have come to accept all that radio technology offers and we take it for granted. By providing a means of sending information without wires it offers the flexibility that is becoming increasingly important in today's highly mobile world.

Radio performs many functions today. One of the most established uses is domestic broadcasting. Most homes have a variety of radio sets, ranging from the relatively simple and cheap portable radios, through the more sophisticated car radios to the high fidelity systems. All of these sets offer a high degree of performance that has resulted from many years of use and development. Even so the performance of radio sets is being improved all the time and new facilities are being introduced. The introduction of wideband FM represented a major improvement when it was first introduced. Stereo facilities are now taken for granted and other enhancements like RDS (Radio Data System) are available in many countries today. Beyond this digital radio or digital audio broadcasting has started. This brings true CD quality to radio broadcasts as well as allowing data to be broadcast at the same time.

Cellular telecommunications is another technology that has been made possible by the development of radio. By the use of radio technology, telecommunications has been given considerably greater flexibility as previously only landlines were available. The flexibility offered by cellular phones has given rise to a phenomenal growth in this area. In

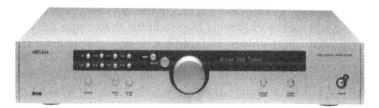

Figure 1.1 *A digital radio tuner (courtesy Arcam)*

Figure 1.2   *An example of a mobile phone handset (courtesy Nokia)*

many countries in the world around 60 to 70 per cent of the population own one of these phones. The enormous usage has enabled a considerable amount of development to be undertaken. The first phones that were used were very large. Now phones can be slipped into a shirt pocket, and all the time the level of functionality they offer is increasing. The first generation systems used analogue technology. The second generation systems use digital technology, and now the third generation systems (3G) are offering high data rate capabilities that will enable them to provide far more facilities and flexibility.

Radio is being used increasingly in many short range 'wireless' applications. Not only are many items such as remote very low power short-range devices including door bells, thermostats and the like being widely used these days, but standards like Bluetooth, HomeRF and 802.11 are being developed and equipment introduced for interconnecting computer devices, mobile phones, and many devices around the home. This is likely to fuel an enormous growth in the short-range radio market as it will mean that offices and the home will use radio instead of wires.

Radio is also used for long-range communications. Satellite technology has enabled radio communications to reliably span the world. Once the short wave bands were used, but even though there are many uses for these frequencies most of the communications we use today are routed via satellites to give communications worldwide. Satellites are used for many purposes. Obviously they enable long-distance communications to be made, but they are also used in applications that include weather monitoring, navigation, search and rescue, direct broadcasting and much more.

Naturally there are many other areas where radio is used today, from point to point communications to other forms of radio links, and it is not possible to mention them all.

There has been a vast amount of investment needed to achieve the high standards of communication needed today. Satellites, microwave links, cellular phone base stations, sophisticated radio data links are but a few that are needed to support today's requirements. To achieve this a number of different types of technology are needed: antennas, transmitters, and receivers as well as several other items are all needed to make up the complete system. This provides a great amount of variety for anyone looking to take radio up as a career or hobby.

Before looking at the emerging technologies and where radio technology is moving, it is always worth having a look at how radio technology arrived at where it is today. In this way it is possible to understand some of the reasons why it is structured and organized in the way it is. The story of the development is fascinating and the very earliest discoveries of major importance can be traced back thousands of years.

The story of radio

The very first discoveries of interest in this story can be traced back to the ancient Greeks and Chinese who investigated the properties of lodestone, a magnetic oxide of iron. In fact the Chinese are often credited with the invention of the magnetic compass. Knowledge of this discovery spread over the known globe and eventually it reached Europe. Here the Greeks with their well-developed society paid most attention to it.

Another early discovery of major importance was that when amber was rubbed against another material, it attracted other objects towards it. This phenomenon possibly added value to the stone when it was traded.

These were only basic discoveries and very little was understood about them. It took many hundreds of years before people started to make any progress.

One of the first people to make anything of these phenomena was William Gilbert, the physician to Queen Elizabeth I of England. He performed a vast number of experiments on magnetism and electrostatics, discovering electrostatic repulsion and the fact that the earth acts as a giant magnet. He published his findings in 1600 in a large volume entitled *De magnete* written in Latin.

After this many other researchers added their own contributions to the newly developing science associated with electricity. Volta, for example, discovered that electricity could be produced by using two dissimilar metals separated by a suitable solution. Ampère made significant advances in linking electricity and magnetism by describing the magnetic forces that exist around a current carrying conductor. Georg Ohm is also well known, especially for determining how current flow was governed in a conductor. This culminated in the famous Ohm's law.

Many other famous names like Faraday, Oersted and Henry made very significant contributions, laying the foundations for developments in radio that were made in the years to follow.

Electricity soon started to find some uses, and became more than just a phenomenon to be demonstrated in the laboratory. The need to communicate swiftly over large distances had long been recognized. The French had set up a system of semaphore towers primarily for military purposes. However, electrical systems were soon introduced. Some early versions were very ambitious. Together Cooke and Wheatsone developed a system of pointers that indicated the required letter. In 1837 the first experimental link was opened between Euston and Camden stations. Although the experiment was a technical success the directors of the company were not impressed and the system was dismantled. Two years later the Great Western Railway Company agreed that a system could be installed between Paddington and West Drayton (not far from where Heathrow Airport stands to the west of London). Two years later this was extended to Slough, allowing communication between the stations. The initial systems required five wires, and as wire production was very expensive it meant that very few systems were installed.

It took a few more years for a system to be developed that would gain widespread acceptance. The inventor was an American artist named Samuel Morse. He had always taken an interest in the new and developing science associated with electricity and while returning by ship from Europe shortly after the discovery of the electromagnet he developed some ideas for an electrical signalling system. On his return to the USA his work commitments took priority and development was particularly slow. Eventually he enlisted assistance from others and the development started to move forwards. After many setbacks he eventually convinced the American Congress to provide funding for a trial line and on 24 May 1844 Morse sent his first message along the line between Baltimore and Washington which read 'What hath God wrought!'. The system quickly gained acceptance because it only used one wire and was easy and reliable to use. Not only did its use spread very rapidly in the USA, but it was used worldwide, linking countries and continents, enabling messages to be sent across the world in a few minutes or hours where they would have taken weeks before.

The Morse system only enabled text messages to be sent. The next major development entailed the sending of sounds over wires. The invention of the telephone is often credited to a Scot named Alexander Graham Bell. He had followed his father in working to enable deaf people to speak and in 1870 the Bell family moved to Canada. Bell was very successful and soon found himself appointed as Professor of Vocal Physiology at Boston University in the USA. His researches involved looking at sound vibrations, and this led him to wonder whether these vibrations could be transmitted along wires in the form of electrical variations. His initial attempts at realizing his idea gave sounds that were unintelligible. Then in 1876 he set up a new system, and the first message that was successfully transmitted over wires was Bell saying 'Mr Watson, come here I want you.' Bell had spilt some acid from a battery over himself and he wanted his assistant Watson to help. In this way the first telephone message was an emergency call!

These were many of the essential foundations that were required before discoveries in radio could begin. Here the first major stone was set in place by a brilliant mathematician and researcher named Maxwell. Born in 1831 he entered Edinburgh University when he was only 16. After graduating, he spent time at a number of universities, but it was when he was at King's College in London that he undertook most of the work into electromagnetic theory for which he is famous. He published three main papers between 1855 and 1864, and finally summarized his work in a book entitled *Treatise on Electricity and Magnetism*. His work proved the existence of an electromagnetic wave. However, much of Maxwell's work was theoretical and he was never able to demonstrate the presence of electromagnetic waves in practice. Sadly, Maxwell died at the early age of 48, and the work he started was left to others to continue.

The quest for the electromagnetic wave took many years. A number of people including Edison and Henry came close to discovering it. However, it was a German named Heinrich Hertz to whom the honour fell.

Hertz performed a wide variety of experiments to prove the existence of these new waves. He also gave a number of demonstrations and lectures. In one of these he used an induction coil connected to a loop of wire in which two large spheres were placed slightly apart. The induction coil generated a large voltage in the circuit causing a spark to jump across the gap. In turn this caused a spark to jump across the gap of a similar coil with two spheres placed within a few metres of the transmitter.

Using other equipment Hertz managed to prove many of the basic properties of these waves. He showed that they had the same velocity as light, and they were refracted and reflected in the same way. As Hertz had discovered the waves, they soon became known as Hertzian waves.

With the existence of the electromagnetic wave firmly established it did not take long before people started to think of using them for communicating. However, to be able to achieve this it was necessary to be able to have a much better way of detecting them. This came in the form of the coherer. A Frenchman named Edouard Branly initially designed it in 1890. He discovered that the resistance of a glass tube filled with metal filings fell from a very high resistance to a few hundred ohms when the filings cohered or clung together when an electrical discharge took place nearby. A small tap on the glass could reset it quite easily.

After its initial discovery the coherer was greatly improved and popularized by an English scientist named Sir Oliver Lodge. Such were his improvements that in 1894 he was able to detect signals from a transmitter a few hundred yards away.

It was around this time that a young Italian named Marconi started to experiment with Hertzian waves. His drive, intuition and business sense enabled this new science to progress much faster than it would otherwise have done.

Marconi was born in Bologna in northern Italy in 1874, receiving a private education during his early years. Despite his parents' expectations he failed to gain a place at Bologna University. Fortunately he had a growing interest in science and a family friend who was a lecturer at the university encouraged him. Marconi was allowed to sit through his lectures and through this he discovered about the new Hertzian waves.

Marconi quickly became interested and soon started to perform many experiments in the attic of his parents' house. Early in the summer of 1895 he managed to transmit a signal a distance of a few yards. By the end of the summer he had succeeded in receiving a signal over 2 km away from the transmitter.

Even at this early stage Marconi was able to see the commercial possibilities. Accordingly he approached the Italian Ministry of Posts with his ideas for wire-less communications, but his proposal was turned down. It was this rejection which caused Marconi to come to England in 1896.

On his arrival he soon set to work and filed a patent for a system of wireless telegraphy using Hertzian waves which was granted on 2 June 1896. In England there was considerably more interest for Marconi's work. Soon he was introduced to a man named William Preece who was the chief engineer of the Post Office.

Marconi gave some preliminary demonstrations of his equipment in the laboratory and then he set up a transmitter and receiver on the roofs of some buildings in London a few hundred yards apart. The success of this demonstration promoted a further demonstration on Salisbury Plain in September 1896 when representatives of the Post Office together with others from the War Office and the Navy were present.

The Navy saw the possibilities of using wireless equipment for communication at sea and they showed considerable interest. However, Marconi also started to sell his equipment to other maritime users. Initially the take-up was slow, but soon other organizations like Lloyd's endorsed its use as a method of sending distress signals and very quickly more vessels were fitted with Marconi's equipment.

Not satisfied with supplying equipment for maritime use, Marconi also started to investigate its use for providing a long-distance communications link. Initial experiments sending a message across the English Channel proved the possibilities and gave valuable propaganda. But the main goal was to be able to send a message across the Atlantic. This was not an easy task. Many difficulties needed to be overcome, but with the help of his team consisting of Vyvyan, Professor Fleming of University College London, Paget, and Kemp, stations at Poldhu in Cornwall and Cape Cod were soon established. Unfortunately the aerials were destroyed in storms and it was decided that their design should be changed. At the same time the site at Cape Cod was abandoned in favour of one at St Johns in Newfoundland. Finally on 12 December 1901 the first transmission was received when the letter 'S' was detected in the receiver and Marconi became a legend in his own time.

This was a major success, and it ably proved the value of wireless as a means of providing long-distance communications. However, it also

Figure 1.3 *Marconi after his transmission (courtesy Marconi plc)*

highlighted the many shortcomings of the systems in use at the time. The most noticeable problem was the lack of sensitivity of the detector in the receiver.

It was this problem that occupied much of the thought of Professor Ambrose Fleming. He had played a significant part in the transatlantic success, designing many parts of the equipment including the transmitter. For a number of years he had known about a phenomenon called the Edison effect. It was found that when a second element was introduced into a light bulb, current would only pass in one direction between the two elements. Both Edison and Fleming had observed the effect, but neither had been able to think of a use for it. However, one day as Fleming was walking down Gower Street in London, he had what he later described as 'a sudden very happy thought'. He realized the possibilities of using the Edison effect in detecting Hertzian waves. Fleming quickly instructed his assistant to set up an experiment to see if the idea worked and to their delight it proved very successful.

Fleming patented the idea on 16 November 1904, calling the device his oscillation valve because of its unidirectional properties with electrical oscillations. In many respects this was the first major component of the electronic era.

The oscillation valve consisted of a filament heated to white heat. This caused electrons to leave the filament because of an effect known as

Figure 1.4   *Fleming's early oscillation valves (courtesy Marconi plc)*

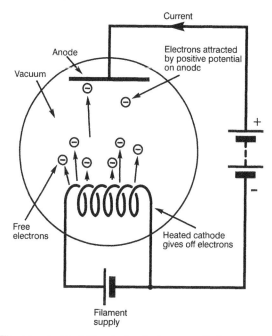

Figure 1.5 *The idea behind Fleming's diode or oscillation valve*

thermionic emission. This makes the filament positively charged and the electrons are attracted back. If another element called an anode is introduced and is given a positive charge then it will attract the electrons away from the heated filament or cathode.

Fleming used this effect to rectify the radio or wireless signals so that the modulation of the alternating current signals was detected. The signals could then be displayed on a meter or heard in headphones.

An American named Lee de Forest made the next major development. He saw the possibilities of Fleming's diode and started to work on copies of the device. In an attempt to be able to circumvent Fleming's patents he investigated a number of other ideas, placing a third electrode into the envelope between the cathode and anode. He called this electrode a grid because it consisted of a wire mesh or grid through which the electronics could pass.

De Forest took out a number of patents for his idea, but it was not until around 1907 that he had developed the new valve, which he called an Audion, sufficiently for it to be used properly. Unknown to him at this time, de Forest had struck on an idea of immense importance, but initially the triode valve was only used as a detector. It took a number of years before it was used as an amplifier.

After their introduction, valves were not widely used. They were expensive to buy, and they were also very expensive to run. At this time valve technology was such that each valve required a separate battery to supply its filament, and as they consumed an amp or more each, the batteries did not last very long. This resulted in other developments coming to the fore. The main one of these was the crystal detector.

These detectors consisted of a crystal that was usually made out of a substance called galena. A thin wire was positioned onto it to make a point contact which acted as a rectifier to detect the signals. These detectors became commonly known as 'cat's whiskers', and were widely used because they were very cheap. Unfortunately they were not very reliable, and often the position of the whisker had to be moved to a better position on the crystal.

It was only after the crystal detector was well established that the Audion was used for amplifying signals. It took until 1911 for it to be used in this role, and de Forest quickly exploited the idea when he started to build amplifiers for use as telephone repeaters.

As the triode valve started to be used as an amplifier its property to act as an oscillator was very quickly discovered. In fact these early valves were often difficult to stop oscillating, especially at radio frequencies. This was put into good use in a number of ways. As far as radio reception was concerned it soon led to the development of the regenerative receiver. Here positive feedback in the circuit was used to increase the sensitivity and selectivity, although the circuit was maintained just before the point of oscillation for the best performance. The American Edwin Armstrong is normally credited with this invention, although others including de Forest, Irving Langmuir in the USA, and Alexander Meissner in Germany also discovered the idea at about the same time.

In 1914, war broke out. With this, wireless research on both sides of the conflict took on a new degree of urgency as its importance was recognized. In developing new and better receivers many famous engineers including H.J. Round, Latour, and Edwin Armstrong on the Allied side and W. Schottky for the Germans devoted vast amounts of research effort into the development of better techniques. Although the idea of the regenerative receiver had given significant improvements over anything that was previously available, receivers still lacked the performance which was sought. Many of the limitations were caused by the poor performance of the valves. High interelectrode capacitances meant that at frequencies above a few hundred kilohertz the valves lacked any usable gain and became very unstable.

H.J. Round made some significant developments in valve technology to try to overcome the problem. However, it was a Frenchman named Lucien Levy who opened the gate to the real solution. In 1917 he devised a system of reception which he claimed would completely eliminate 'parasites and ordinary interference'. To achieve this he used the principle of beats to convert signals from a high frequency down to a lower one where the variable filters could be made more selective.

The American Edwin Armstrong later used the same basic principle in his receiver. He used a variable frequency local oscillator to convert the incoming signals down in frequency. A fixed frequency filter was then used to give the selectivity. By varying the frequency of the oscillator the receiver could be tuned from one frequency to the next. This idea had a number of advantages. Using a low frequency for the filter gave better selectivity, and by keeping its frequency fixed, several stages could be placed in series to give even better performance. He also included a significant amount of amplification at this lower intermediate frequency stage, where better gain could be achieved from the valves.

Armstrong patented his idea for the receiver, he called a supersonic heterodyne, or superhet receiver, on 30 December 1918. It was too late to see any real use in the war and the superhet did not see much use for a number of years. The main reason for this was that they used more valves than their regenerative counterparts, and for normal broadcast reception the additional performance was not needed.

At this time most long-distance communications were undertaken on very long wavelengths. As a result amateur experiments were relegated to the short wave bands with wavelengths shorter than 200 metres. These bands were thought to be of little use, but undeterred by this, radio amateurs sought to disprove this and make long-distance contacts.

In America there were a large number of radio amateurs who were permitted to use relatively high powers, and reports of long-distance contacts soon started to appear. With this in mind, the possibility of short wave communication across the Atlantic was thought to be a possibility. The first attempts were made in December 1921, with British stations listening for their American counterparts who could use much higher powers. These first attempts proved to be unsuccessful; however, the following winter many stations were heard. Then in November 1923, the first two way contact was made between an American in Connecticut and a French station. With this barrier broken regular transatlantic communications started to be made. Contacts were also made over further distances, one between London and Dunedin in New Zealand was made in October 1924.

While radio amateurs took the limelight, commercial interests were also investigating the possibilities of the short wave bands. A telephone circuit was set up between Hendon in North London and Birmingham in

1921, and later that year another was inaugurated between Southwold in Sussex and Zandvoort in Holland.

Seeing the success of all these projects Marconi decided to make further investigations into the properties of these bands. He set up a 12 kW transmitting station at Poldhu, the site of his long wave transmitter in 1901, and he used receivers on his yacht *Elettra* to monitor the transmissions. He found that signals up to St Vincent in the West Indies (a distance of 2300 miles) could be received at sufficient strength to support a telephone channel.

Many other tests and experiments were performed, and slowly the short wave bands started to be used for international traffic. By the late 1920s radio was providing a valuable method of providing international telephone circuits. Links were even opened up to ocean liners to provide instant communication for passengers who were travelling on board.

Wireless was not only used for the transmission of commercial traffic. Some early broadcasts of entertainment were made, with the Marconi station in Chelmsford making the first regular broadcasts in the UK. This station opened on 23 February 1920, making a daily broadcast of half an hour of speech, music and news. About four months later on 20 June history was made when the Australian singer Dame Nellie Melba made the first broadcast by a professional artist. The transmission made the front pages of the newspapers. However, interference caused by the transmissions forced the closure of the station.

Public opinion soon meant that the decision had to be reversed. On 14 February 1922 the Marconi station licensed as 2MT was allowed to broadcast a 15 minute programme. This proved to be very successful and on 11 May a station commenced broadcasting from Marconi's headquarters in the centre of London. This was the famous station 2LO, which was taken over by the British Broadcasting Company (later to become the British Broadcasting Corporation or BBC) in November of that year.

With broadcasting now established more stations were set up. Before long most areas of the British Isles had their own station. Broadcasting increased in other countries as well. In the USA the rise in the number of stations was far greater, placing far greater requirements on receiving equipment, and encouraging the use of Armstrong's superhet receiver.

The short wavebands also started to be used for international broadcasting. An American station owned by the Westinghouse Company undertook early experiments. Transmitting from Pittsburg on a wavelength of 62.5 metres, the station (KDKA) was successfully received in Britain and actually rebroadcast. In 1927 an experimental short wave broadcasting service was set up in Britain. The first station was located in Chelmsford and this service continued until 1932 when the BBC Empire Service transmitting from a site in Daventry was opened. Its aim was to keep all parts of the British Empire in contact with London by being able

Figure 1.6   *A superhet radio receiver from the early 1930s*

to receive at least one transmission a day. This was the forerunner of today's BBC World Service.

With short wave broadcasting well established attention was soon turned to the possibility of using ultra short waves for communication. Initially this posed problems with suitable devices for operation at these frequencies. However, these problems were slowly overcome and in March 1931 Standard Telephones and Cables Limited gave a demonstration of telephony across the Straits of Dover using a wavelength of just 18 cm. The British Post Office also installed some services on these frequencies. In 1932 they opened a telephone circuit across the Severn on 5 metres which saved having to use a much longer cable route through the Severn Tunnel.

The idea of modulating the frequency of a signal instead of its amplitude had been known from the earliest days of radio. Early experiments had been undertaken in 1902 to investigate its viability. Later ones as well had treated it in a similar way to AM. In attempts to reduce the levels of background noise and interference the bandwidth had been reduced and this had only served to impair its performance.

However, Edwin Armstrong, who was already a millionaire from his previous inventions, turned traditional thinking around. Instead of reducing the bandwidth, he increased it. Early interest in the idea was very limited and so Armstrong set up his own transmitter and receiver to demonstrate the idea in 1935. Using a frequency of 110 MHz he was able to conclusively prove that wideband FM gave significant improvements over traditional AM systems for high quality transmissions. Also using higher frequencies he would be able to reduce the congestion on the long and medium wave bands which were already overcrowded.

Despite the success of the demonstration, interest from broadcasters was very limited. Undeterred by this, Armstrong set up his own station that opened in July 1939. Soon other stations saw the possibilities of improved reception and within six months over 140 applications had been submitted in the USA for new VHF FM stations.

The Second World War delayed the spread of wideband FM, and it was not until 1954 that, the BBC launched its first FM service in the UK. Since then the growth of wideband FM has been enormous. It now attracts greater audiences than the traditional AM broadcasts because of its improved sound reproduction and resilience to noise and fading.

Communications will always be a very important use for radio. During 1946, the scientist and science fiction writer Arthur C. Clarke published a revolutionary article in the magazine *Wireless World*. In this he proposed the use of three orbiting satellites to give worldwide communication. While his ideas were remarkably sound, the technology did not exist to put them into practice. However, during the Second World War the Germans had made major advances into rocket technology. These were the first steps that enabled rockets to put satellites into space, and during the 1950s many further advances were made.

Many of the first attempts at putting satellites into space failed dramatically. However, on 4 October 1957 the Russians succeeded in placing a football-sized satellite into a low orbit. Named Sputnik 1 it radiated a distinctive bleep that was heard by many professional and amateur listeners all over the world.

Sputnik was the first major milestone in the satellite age. Soon others followed. In 1960 America launched Echo, a 33 metre diameter aluminium coated balloon. This could easily be seen from earth as it passed overhead and it enabled radio signals to be reflected to provide extended distances for radio communications. This was only the first step in making a proper communications satellite. In 1962 Telstar was launched. This was a private venture funded by AT&T and on 23 July 1962 it succeeded in providing the first live transatlantic pictures relayed by satellite.

After the success of Telstar developments progressed at a rapid rate and many more satellites were launched. Soon they were providing an enormous variety of services from communications to weather monitoring, and direct broadcasting to navigation.

Satellites were by no means the only improvement for communications. With electronic technology progressing at an ever increasing rate the need for personal communications was seen as the next major goal. As a result the idea of a truly mobile phone developed. The basic idea appeared at Bell Laboratories in the USA in the early 1970s but they were

Figure 1.7   *Telstar (courtesy NASA)*

not available for use until the 1980s. In fact the first systems were introduced in the UK in 1984. During the 1980s these first generation analogue systems experienced rapid growth. In view of the limitations of the analogue systems digital systems were envisaged. The most widely used digital system, GSM, was first introduced for commercial use in 1991 and its use spread rapidly. The GSM system was intended as a pan-European system, although its use spread beyond Europe. The acronym GSM initially stood for Groupe Spécial Mobile, but later this changed to Global System for Mobile Communications. In the USA the first CDMA system was launched with the introduction of cdmaOne. With the success of the second generation systems, third generation systems capable of high data rate transmissions were devised.

Since the first radio or wireless transmissions were made the field of radio has grown by enormous proportions. Now, more than ever, it is a cornerstone to our everyday life, both at home and at work. With technology progressing as it is, radio is likely to become even more important in the years to come, giving more flexibility and higher quality services. It is also likely to find an even greater variety of uses, many of which have not even been conceived yet.

Radio tomorrow

There is a very large amount of development being undertaken in a wide variety of areas associated with radio. Today people require greater degrees of freedom and flexibility, and radio and wireless links are one way in which this can be provided. Fast and flexible communication is also required across vast distances, and again radio provides a cost effective means of providing this. In view of these and other requirements radio is one of the fastest growing areas of technology today.

Since their real introduction in the 1980s cellular telecommunications has experienced a phenomenal rate of growth. In view of their cost they were initially only used by businessmen. However, as costs fell their use increased with market penetration reaching levels of 70 per cent in some countries. Now with the new technologies becoming available there is a growing emphasis on using cellular technology for data transmission and remote working.

New short-range wireless technologies are becoming far more widely used. Not only are items like wireless controlled door bells, heating thermostats and the like becoming commonplace around the home, but new technologies are enabling items like computer peripherals to talk to one another without the need for many interconnecting wires to be used. This could enable connecting computers together to be far easier than at present.

With digital radio now established, this type of technology is likely to grow. Not only does this provide a much greater level of quality than is possible on FM, but it enables many more services to be provided giving far more flexibility to the medium.

Apart from introducing new services, many other radio developments are focused on pushing forwards to boundaries of technology. The pressure on the amount of radio spectrum that is available means that higher frequencies need to be used. To meet this need new semiconductor devices are being developed which are capable of operating at frequencies much higher in frequency than before. Although these devices may not be widely available for a number of years, devices which were at the forefront of technology some years ago are now in widespread use enabling many of the systems which we take for granted to function.

These developments represent just a few of the areas where radio technology is moving forwards. However, they are not the only areas. In view of the importance of radio in today's society, there are many areas where new developments are being sought to enable new radio applications to be implemented and others to be able to operate it more effectively. All of them are an indication of the importance that the electronics industry attaches to radio. This means that it is a technology for the future.

2 Radio waves and propagation

The properties of radio waves and the way in which they travel or propagate are of prime importance in the study of radio technology. These waves can travel over vast distances enabling communication to be made where no other means is possible. Using them communication can be established over distances ranging from a few metres to many thousands of miles. This enables telephone conversations and many other forms of communication to be made with people on the other side of the world using short wave propagation or satellites. Radio waves can be received over even greater distances. Radio telescopes pick up minute signals from sources many light years away.

Radio waves are a form of radiation known as electromagnetic waves. As they contain both electric and magnetic elements it is first necessary to take a look at these fields before looking at the electromagnetic wave itself.

Electric fields

Any electrically charged object whether it has a static charge or is carrying a current has an electric field associated with it. It is a commonly known fact that like charges repel one another and opposite charges attract. This can be demonstrated in a number of ways. Hair often tends to stand up after it has been brushed or combed. The brushing action generates an electrostatic charge on the hairs, and as they all have the same type of charge they tend to repel one another and stand up. In this way it can be seen that a force is exerted between them. Examples like these are quite dramatic and result because the voltages that are involved are very high and can typically be many kilovolts. However, even the comparatively low voltages that are found in electronic circuits exhibit the same effects although to a much smaller degree.

The electric field radiates out from any item with an electric potential as shown in Figure 2.1. The electrostatic potential falls away as the distance from the object is increased. Take the example of a charged sphere with a potential of 10 volts. At the surface of the sphere the

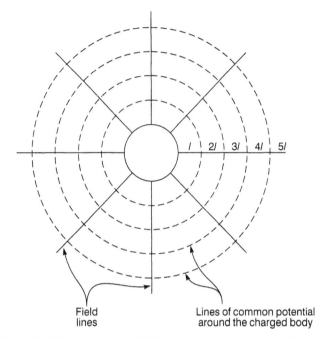

Field
lines

Lines of common potential
around the charged body

Figure 2.1   *Field lines and potential lines around a charged sphere*

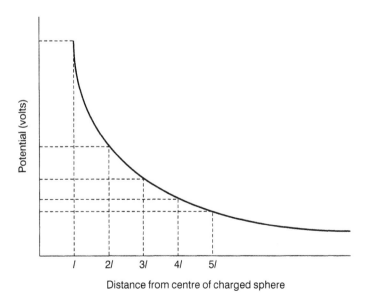

Distance from centre of charged sphere

Figure 2.2   *Variation of potential with distance from the charged sphere*

electrostatic potential is 10 volts. However, as the distance from the sphere is increased, this potential starts to fall. It can be seen that it is possible to draw lines of equal potential around the sphere as shown in Figure 2.1.

The potential falls away as the distance is increased from the sphere. It can be shown that the potential falls away as the inverse of the distance, i.e. doubling the distance halves the potential. The variation of potential with the distance from the sphere is shown in Figure 2.2.

The electric field gives the direction and magnitude of the force on a charged object. The field intensity is the negative value of the slope in Figure 2.2. The slope of a curve plotted on a graph is the rate of change of a variable. In this case it represents the rate of change of the potential with distance at a particular point. This is known as the potential gradient. It is found that the potential gradient varies as the inverse square of the distance. In other words doubling the distance reduces the potential gradient by a factor of four.

Magnetic fields

Magnetic fields are also important. Like electric charges, magnets attract and repel one another. Analogous to the positive and negative charges, magnets have two types of pole, namely a north and a south pole. Like poles repel and dissimilar ones attract. In the case of magnets it is also found that the magnetic field strength falls away. It is found that it falls away as the inverse square of the distance.

While the first magnets to be used were permanent magnets, much later it was found that an electric current flowing in a conductor

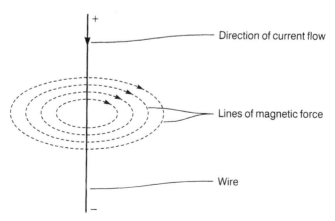

Figure 2.3 *Lines of magnetic force around a current carrying conductor*

generated a magnetic field. This could be detected by the fact that a compass needle placed close to the conductor would deflect. The lines of force are in a particular direction around the wire as shown. An easy method of determining which way they go around the conductor is to use the corkscrew rule. Imagine a right-handed corkscrew being driven into a cork on the direction of the current flow. The lines of force will be in the direction of rotation of the corkscrew.

Radio waves

As already mentioned radio signals are a form of electromagnetic wave. They consist of the same basic type of radiation as light, ultraviolet and infrared rays, differing from them in their wavelength and frequency. These waves are quite complicated in their make-up, having both electric and magnetic components that are inseparable. The planes of these fields are at right angles to one another and to the direction of motion of the wave. These waves can be visualized as shown in Figure 2.4.

The electric field results from the voltage changes occurring in the antenna which is radiating the signal, and the magnetic field changes result from the current flow. It is also found that the lines of force in the electric field run along the same axis as the antenna, but spreading out as they move away from it. This electric field is measured in terms of the

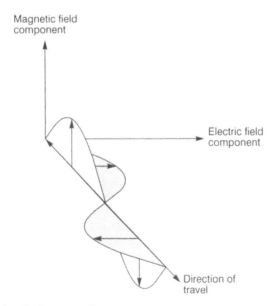

Figure 2.4   *An electromagnetic wave*

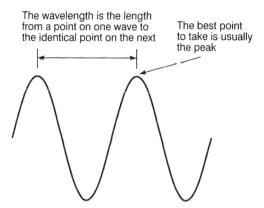

The wavelength is the length from a point on one wave to the identical point on the next

The best point to take is usually the peak

Figure 2.5 *The wavelength of an electromagnetic wave*

change of potential over a given distance, e.g. volts per metre, and this is known as the field strength.

There are a number of properties of a wave. The first is its wavelength. This is the distance between a point on one wave to the identical point on the next as shown in Figure 2.5. One of the most obvious points to choose is the peak as this can be easily identified although any point is acceptable.

The second property of the electromagnetic wave is its frequency. This is the number of times a particular point on the wave moves up and down in a given time (normally a second). The unit of frequency is the hertz and it is equal to one cycle per second. This unit is named after the German scientist who discovered radio waves. The frequencies used in radio are usually very high. Accordingly the prefixes kilo, mega, and giga are often seen. 1 kHz is 1000 Hz, 1 MHz is a million hertz, and 1 GHz is a thousand million hertz, i.e. 1000 MHz. Originally the unit of frequency was not given a name and cycles per second (c/s) were used. Some older books may show these units together with their prefixes: kc/s, Mc/s, etc. for higher frequencies.

The third major property of the wave is its velocity. Radio waves travel at the same speed as light. For most practical purposes the speed is taken to be 300 000 000 metres per second although a more exact value is 299 792 500 metres per second.

Frequency to wavelength conversion

Many years ago the position of stations on the radio dial was given in terms of wavelengths. A station might have had a wavelength of 1500 metres. Today stations give out their frequency because nowadays this is far easier to measure. A frequency counter can be used to measure this very accurately, and with today's technology their cost is relatively low. It is very easy to relate the frequency and wavelength as they are linked by the speed of light as shown:

$$\lambda = \frac{c}{f}$$

where λ = the wavelength in metres
   f = frequency in hertz
   c = speed of radio waves (light) taken as 300 000 000 metres per
        second for all practical purposes

Taking the previous example the wavelength of 1500 metres corresponds to a frequency of 300 000 000/1500 or 200 thousand hertz (200 kHz).

Radio spectrum

Electromagnetic waves have a wide variety of frequencies. Radio signals have the lowest frequency, and hence the longest wavelengths. Above the radio spectrum, other forms of radiation can be found. These include infrared radiation, light, ultraviolet and a number of other forms of radiation as shown in Figure 2.6.

The radio spectrum itself covers an enormous range. At the bottom end of the spectrum there are signals of just a few kilohertz, whereas at the top end new semiconductor devices are being developed which operate at frequencies of a hundred GHz and more. Between these extremes lie all the signals with which we are familiar. It can be seen that there is a vast amount of spectrum space available for transmissions. To make it easy to refer to different portions of the spectrum,

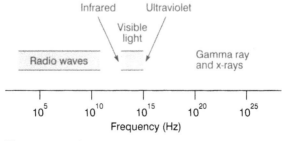

Figure 2.6   *Electromagnetic wave spectrum*

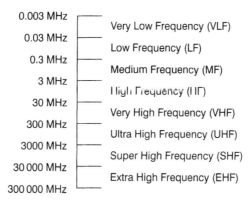

Figure 2.7 *The radio spectrum*

designations are given to them as shown in Figure 2.7. It can be seen from this that transmissions in the long wave broadcast band (140.5 to 283.5 kHz) available in some parts of the world fall into the low frequency or LF portion of the spectrum along with navigational beacons and many other types of transmission.

Moving up in frequency, the medium frequency or MF section of the spectrum can be found. The medium wave broadcast band can be found here. Above this are the lowest frequencies of what may be considered to be the short wave bands. A number of users including maritime communications and a 'tropical' broadcast band can be found.

Between 3 and 30 MHz is the high frequency or HF portion. Within this frequency range lie the real short wave bands. Signals from all over the world can be heard. Broadcasters, maritime, military, weather information, radio amateurs, news links and a host of other general point to point communications fill the bands.

Moving up further the very high frequency or VHF part of the spectrum is encountered. This contains a large number of mobile users. 'Radio Taxis' and the like have allocations here, as do the familiar VHF FM broadcasts.

In the ultra high frequency or UHF part of the spectrum most of the terrestrial television stations are located. In addition to these there are more mobile users including cellular telephones generally around 850 and 900 MHz as well as 1800 and 1900 MHz dependent upon the country.

Above this in the super high frequency or SHF and extremely high frequency or EHF portions of the spectrum there are many uses for the radio spectrum. They are being used increasingly for commercial satellite and point to point communications.

Polarization

There are many characteristics of electromagnetic waves. One is polarization. Broadly speaking the polarization indicates the plane in which the
wave is vibrating. In view of the fact that electromagnetic waves consist
of electric and magnetic components in different planes, it is necessary to
define a convention. Accordingly the polarization plane is taken to be that
of the electric component.

The polarization of a radio wave can be very important because
antennas are sensitive to polarization, and generally only receive or
transmit a signal with a particular polarization. For most antennas it is
very easy to determine the polarization. It is simply in the same plane as
the elements of the antenna. So a vertical antenna (i.e. one with vertical
elements) will receive vertically polarized signals best and similarly a
horizontal antenna will receive horizontally polarized signals.

Vertical and horizontal are the simplest forms of polarization and they
both fall into a category known as linear polarization. However, it is also
possible to use circular polarization. This has a number of benefits for
areas such as satellite applications where it helps overcome the effects of
propagation anomalies, ground reflections and the effects of the spin that
occur on many satellites. Circular polarization is a little more difficult to
visualize than linear polarization. However, it can be imagined by
visualizing a signal propagating from an antenna that is rotating. The tip
of the electric field vector will then be seen to trace out a helix or
corkscrew as it travels away from the antenna. Circular polarization can
be seen to be either right or left handed dependent upon the direction of
rotation as seen from the transmitter.

Another form of polarization is known as elliptical polarization. It
occurs when there is a mix of linear and circular polarization. This can be
visualized as before by the tip of the electric field vector tracing out an
elliptically shaped corkscrew.

It can be seen that as an antenna transmits and receives a signal with a
certain polarization that the polarization of the transmitting and receiving
antennas is important. This is particularly true in free space, because once
a signal has been transmitted its polarization will remain the same. In
order to receive the maximum signal both transmitting and receiving
antennas must be in the same plane. If for any reason their polarizations
are at right angles to one another (i.e. cross-polarized) then in theory no
signal would be received. A similar situation exists for circular polarization. A right-handed circularly polarized antenna will not receive a left-
hand polarized signal. However, a linearly polarized antenna will be able
to receive a circularly polarized signal. The strength will be equal
whether the antenna is mounted vertically, horizontally or in any other
plane at right angles to the incoming signal, but the signal level will be

3 dB less than if a circularly polarized antenna of the same sense was used.

For terrestrial applications it is found that once a signal has been transmitted then its polarization will remain broadly the same. However, reflections from objects in the path can change the polarization. As the received signal is the sum of the direct signal plus a number of reflected signals the overall polarization of the signal can change slightly although it remains broadly the same.

Different types of polarization are used in different applications to enable their advantages to be used. Linear polarization is by far the most widely used. Vertical polarization is often used for mobile or point to point applications. This is because many vertical antennas have an omnidirectional radiation pattern and this means that the antennas do not have to be reorientated as positions are changed if, for example, a vehicle in which a transmitter or receiver is moved. For other applications the polarization is often determined by antenna considerations. Some large multi-element antenna arrays can be mounted in a horizontal plane more easily than in the vertical plane and this determines the standard polarization in many cases. However, for some applications there are marginal differences between horizontal and vertical polarization. For example, medium wave broadcast stations generally use vertical polarization because propagation over the earth is marginally better using vertical polarization, whereas horizontal polarization shows a marginal improvement for long-distance communications using the ionosphere. Circular polarization is sometimes used for satellite communications as there are some advantages in terms of propagation and in overcoming the fading caused if the satellite is changing its orientation.

How radio signals travel

Radio signals are very similar to light waves and behave in a very similar way. Obviously there are some differences caused by the enormous variation in frequency between the two, but in essence they are the same.

A signal may be radiated or transmitted at a certain point, and the radio waves travel outwards, much like the waves seen on a pond if a stone is dropped into it. As they move outwards they become weaker as they have to cover a much wider area. However, they can still travel over enormous distances. Light can be seen from stars many light years away. Radio waves can also travel over similar distances. As distant galaxies and quasars emit radio signals these can be detected by radio telescopes which can pick up the minute signals and then analyse them to give us further clues about what exists in the outer extremities of the universe.

The loss of a signal travelling in free space can easily be determined as the only two variables are the distance and the frequency in use. The distance is the straight line distance between the transmitter and receiver. The loss resulting from the frequency in use arises from the fact that at higher frequencies the antennas are smaller and hence the received signal is smaller.

$$\text{Loss (dB)} = 32.45 + 20 \log_{10} \text{(frequency in MHz)} + 20 \log_{10} \text{(distance in km)}$$

Using the figure for a loss in a system it is quite easy to calculate the receiver levels for a given transmitter power. Transmitter and receiver power levels should both be expressed in dBW (dB relative to one watt) or dBm (dB relative to a milliwatt). The antenna gain naturally has an effect and gain levels are expressed relative to an isotropic radiator, i.e. one that radiates equally in all directions. Further details of antenna gains and isotropic radiators are given in Chapter 4. Feeder losses should also be taken into account as these have an effect on the overall signal levels and may be significant in some instances.

$$P_r(\text{dBm}) = P_t(\text{dBm}) + G_{ta}(\text{dB}) - L_{tf}(\text{dB}) - L_{path} + G_{ra}(\text{dB}) - L_{rf}(\text{dB})$$

where
P_r = received power level
P_t = transmitter power level
G_{ta} = gain of the transmitter antenna
L_{tf} = loss of the transmitter feeder
L_{path} = path loss
G_{ra} = receiver antenna gain
L_{rf} = loss of the receiver feeder

Refraction, reflection and diffraction

In the same way that light waves can be reflected by a mirror, so radio waves can also be reflected. When this occurs, the angle of incidence is equal to the angle of reflection for a conducting surface as would be expected for light. When a signal is reflected there is normally some loss of the signal, either through absorption, or as a result of some of the signal passing into the medium. For radio signals surfaces such as the sea provide good reflecting surfaces, whereas desert areas are poor reflectors.

Refraction of radio waves is obviously very similar to that of light. It occurs as the wave passes through areas where the refractive index changes. For light waves this can be demonstrated by placing one end of

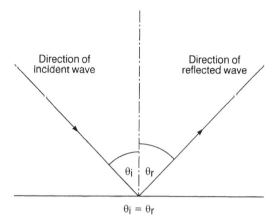

Figure 2.8   *Reflection of an electromagnetic wave*

a stick into some water. It appears that the section of stick entering the water is bent. This occurs because the direction of the light changes as it moves from an area of one refractive index to another. The same is true for radio waves. In fact the angle of incidence and the angle of refraction are linked by Snell's law that states:

$$\mu_1 \sin \theta_1 = \mu_2 \sin \theta_2$$

In many cases where radio waves are travelling through the atmosphere there is a gradual change in the refractive index of the medium. This causes a steady bending of the wave rather than an immediate change in direction.

Diffraction is another phenomenon that affects radio waves and light waves alike. It is found that when signals encounter an obstacle they tend to travel around them as shown in Figure 2.10. The effect can be explained by Huygen's principle. This states that each point on a spherical wave front can be considered as a source of a secondary wave front. Even though there will be a shadow zone immediately behind the obstacle, the signal will diffract around the obstacle and start to fill the void, thereby enabling reception behind the obstacle even though it is not in the direct line of sight of the transmitter. It is found that diffraction is more pronounced when the obstacle approaches a 'knife edge'. A mountain ridge may provide a sufficiently sharp edge. A more rounded obstacle will not produce such a marked effect. It is also found that low frequency signals diffract more markedly than higher frequency ones. Thus signals on the long wave band are able to provide coverage even in hilly or mountainous terrain where signals at VHF and higher would not.

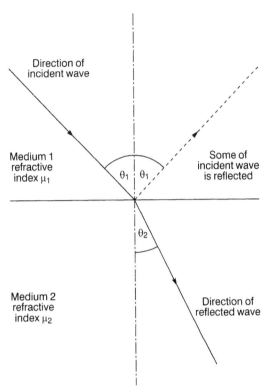

Figure 2.9 *Refraction of an electromagnetic wave at the boundary between two areas of differing refractive index*

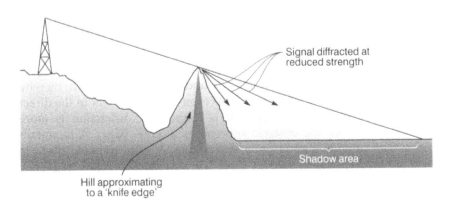

Figure 2.10 *Diffraction of a radio signal around an obstacle*

Reflected signals

Signals that travel near to other objects suffer reflections from a variety of objects. One is the earth itself, but others may be local buildings, or in fact anything that can reflect or partially reflect radio waves. As a result the received signal is the sum of a variety of signals from the transmitter that have reached the receiving antenna via a variety of paths. Each will have a slightly different path length and this will mean that the signals will not reach the receiver with the same phase. As a result some will reinforce the strength of the overall signal while others will interfere and reduce the overall level. This effect can be noticed when an aircraft flies overhead and the overall strength of a signal varies as the aircraft moves and the path length of the signal reflected from it changes. This causes the signal to flutter.

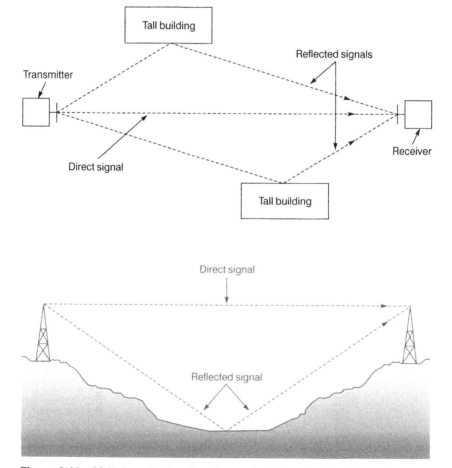

Figure 2.11 *Multiple paths lengths of a received signal arising form reflections*

Not only are signals reflected by visible objects, but areas like the atmosphere have a significant effect on signals, reflecting and refracting them, and enabling them to travel over distances well beyond the line of sight. Before investigating the different ways in which this can happen it is first necessary to take a look at the atmosphere where these effects occur and investigate its make-up.

Layers above the earth

The atmosphere above the earth consists of many layers as shown in Figure 2.12. Some of them have a considerable affect on radio waves whereas others do not. Closest to the surface is the troposphere. This region has very little effect on short wave frequencies below 30 MHz, although at frequencies above this it plays a major role. At certain times transmission distances may be increased from a few tens of kilometres to a few hundred kilometres. This is the area that governs the weather, and in view of this the weather and radio propagation at these frequencies is closely linked.

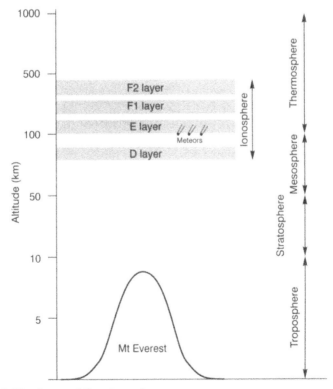

Figure 2.12   *Areas of the atmosphere*

Above the troposphere, the stratosphere is to be found. This has little effect on radio waves, but above it in the mesosphere and thermosphere the levels of ionization rise in what is collectively called the ionosphere.

The ionosphere is formed as the result of a complicated process where the solar radiation together with solar and to a minor degree cosmic particles affect the atmosphere. This causes some of the air molecules to ionize, forming free electrons and positively charged ions. As the air in these areas is relatively sparse it takes some time for them to recombine. These free electrons affect radio waves causing them to be attenuated or bent back towards the earth.

The level of ionization starts to rise above altitudes of 30 km, but there are areas where the density is higher, giving the appearance of layers when viewed by their effect on radio waves. These layers have been designated the letters D, E, and F to identify them. There is also a C layer below the D layer but its level of ionization is very low and it has no noticeable effect on radio waves.

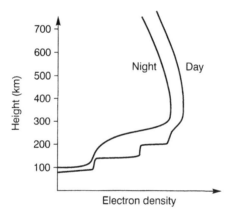

Figure 2.13 *Approximate ionization levels above the earth*

The degree of ionization varies with time, and is dependent upon the amount of radiation received from the sun. At night when the layers are hidden from the sun, the level of ionization falls. Some layers disappear while others are greatly reduced in intensity.

Other factors influence the level of ionization. One is the season of the year. In the same way that more heat is received from the sun in summer, so too the amount of radiation received by the upper atmosphere is increased. Similarly the amount of radiation received in winter is less.

The number of sunspots on the sun has a major affect on the ionosphere. These spots indicate areas of very high magnetic fields. It is found that the number of spots varies very considerably. They have been monitored for over 200 years and it has been found that the number varies over a cycle of approximately 11 years. This figure is an average, and any particular cycle may vary in length from about nine to 13 years. At the peak of the cycle there may be as many as 200 spots while at the minimum the number may be in single figures, and on occasions none have been detected.

Under no circumstances should the sun be viewed directly, or even through dark sunglasses. This is very dangerous and people have lost the sight of an eye trying it.

Sunspots affect radio propagation because they emit vast amounts of radiation. In turn this increases the level of ionization in the ionosphere. Accordingly radio propagation varies in line with the sunspot cycle.

Each of the bands or layers in the ionosphere acts in a slightly different way, affecting different frequencies. The lowest layer is the D layer at a height of around 75 km. Instead of reflecting signals, this layer tends to absorb any signals that it affects. The reason for this is that the air density is very much greater at its altitude and power is absorbed when the electrons are excited. However, this layer only affects signals up to about two megahertz or so. It is for this reason that only local ground wave signals are heard on the medium wave broadcast band during the day.

The D layer has a relatively low electron density and levels of ionization fall relatively quickly. As a result it is only present when

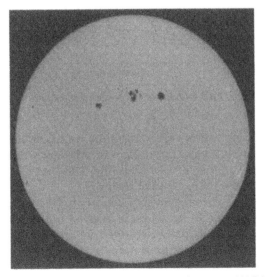

Figure 2.14   *Sunspots on the surface of the sun (courtesy NASA)*

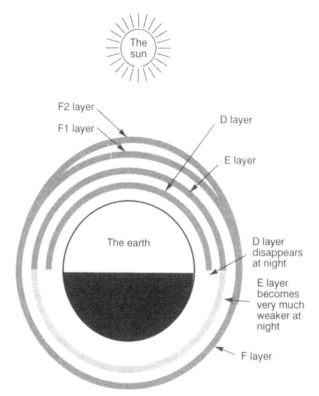

Figure 2.15   *Variation of the ionized layers during the day*

radiation is being received from the sun. This means that it is much weaker in the evening and not present at night. When this happens it means that low frequency signals can be reflected by higher layers. This is why signals from much further afield can be heard on the medium wave band at night.

Above the D layer the E layer is found. At a height of around 110 km, this layer has a higher level of ionization than the D layer. It reflects or more correctly refracts the signals that reach it, rather than absorbing them. However, there is a degree of attenuation with any signal reflected by the ionosphere. The atmosphere is still relatively dense at the altitude of the E layer. This means that the ions recombine quite quickly and levels of ionization sufficient to reflect radio waves are only present during the hours of daylight. After sunset the number of free ions falls relatively quickly to a level where they usually have little effect on radio waves.

The F layer is found at heights between 200 and 400 km. Like the E layer it tends to reflect signals that reach it. It has the highest level of ionization, being the most exposed to the sun's radiation. During the course of the day the level of ionization changes quite significantly. This results in it often splitting into two distinct layers during the day. The lower one called the F1 layer is found at a height of around 200 km, then at a height of between 300 and 400 km there is the F2 layer. At night when the F layer becomes a single layer its height is around 250 km. The levels of ionization fall as the night progresses, reaching a minimum around sunrise. At this time levels of ionization start to rise again.

Often it is easy to consider the ionosphere as a number of fixed layers. However, it should be remembered that it is not a perfect 'reflector'. The various layers do not have defined boundaries and the overall state of the ionosphere is always changing. This means that it is not easy to state exact hard and fast rules for many of its attributes.

Ground wave

The signal can propagate over the reception area in a number of ways. The ground wave is the way by which signals in the long and medium wave bands are generally heard.

When a signal is transmitted from an antenna it spreads out, and can be picked up by receivers that are in the line of sight. Signals on frequencies in the long and medium wave bands (i.e. LF and MF bands) can be received over greater distances than this. This happens because the signals tend to follow the earth's curvature using what is termed the ground wave. It occurs because currents are induced in the earth's surface. This slows the wave front down nearest to the earth causing it to

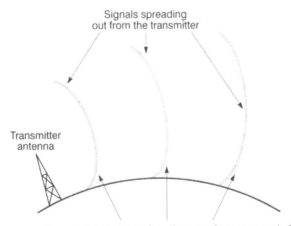

Figure 2.16   *A ground wave*

tilt downwards, and enabling it to follow the curvature, travelling over distances that are well beyond the horizon.

The ground wave is generally only used for signals below about 2 MHz. It is found that as the frequency increases the attenuation of the whole signal increases and the coverage is considerably reduced. Obviously the exact range will depend on many factors. Typically a high power medium wave station may be heard over distances of a 150 km and more. There are also many low power broadcast stations running 100 watts or so. These might have a coverage area extending to 15 or 20 miles.

As the affects of attenuation increase with frequency, even very high power short wave stations are only heard over relatively short distances using ground wave. Instead these stations use reflections from layers high up in the atmosphere to achieve coverage to areas all over the world.

Skywaves

Radio signals travelling away from the earth's surface are called skywaves and they reach the layers of the ionosphere. Here they may be absorbed, reflected back to earth or they may pass straight through into outer space. If they are reflected the signals will be heard over distances which are many times the line of sight. An exact explanation of the way in which the ionization in the atmosphere affects radio waves is very complicated. However, it is possible to gain an understanding of the basic concepts from a simpler explanation.

Basically the radio waves enter the layer of increasing ionization, and as they do so the ionization acts on the signal, bending it or refracting it back towards the area of lesser ionization. To the observer it appears that the radio wave has been reflected by the ionosphere.

When the signal reaches the ionization, it sets the free electrons in motion and they act as if they formed millions of minute antennas. The electrons retransmit the signal, but with a slightly different phase. This has the result that the signal is made to bend away from the area of higher electron density. As the density of electrons increases as the signal enters the layer, the signal is bent back towards the surface of the earth, so that it can often be received many thousands of kilometres away from where it was transmitted.

The effect is very dependent upon the electron density and the frequency. As frequencies increase much higher electron densities are required to give the same degree of refraction.

The way in which radio waves travel through the ionosphere, are absorbed, reflected or pass straight through is dependent upon the

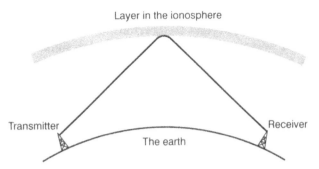

Figure 2.17   *Signals reflected and returned to earth by the ionosphere*

frequency in use. Low frequency signals are affected in totally different ways to those at the top end of the short wave spectrum. This is borne out by the fact that medium wave signals are heard over relatively short distances, and at higher frequencies signals from much further afield can be heard. It may also be found that on frequencies at the top end of the short wave spectrum, no signals may be heard on some days.

To explain how the effects change with frequency take the example of a low frequency signal transmitting in the medium wave band at a frequency of f_1. The signal spreads out in all directions along the earth's surface as a ground wave that is picked up over the service area. Some radiation also travels up to the ionosphere. However, because of the frequency in use the D layer absorbs the signal. At night the D layer disappears and the signals can then pass on being reflected by the higher layers.

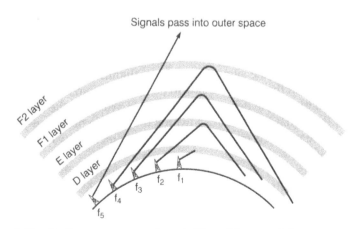

Figure 2.18   *Radio wave propagation at different frequencies*

Signals higher in frequency at f_2 pass straight through the D layer. When they reach the E layer they can be affected by it being reflected back to earth. The frequency at which signals start to penetrate the D layer in the day is difficult to define as it changes with a variety of factors including the level of ionization and angle of incidence. However, it is often in the region of 2 MHz or 3 MHz.

Also as the frequency increases the ground wave coverage decreases. Medium wave broadcast stations can be heard over distances of many tens of miles. At frequencies in the short wave bands this is much smaller. Above 10 MHz signals may only be heard over a few kilometres, dependent upon the power and antennas being used.

The E layer only tends to reflect signals in the lower part of the short wave spectrum to earth. As the frequency increases, signals penetrate further into the layer, eventually passing right through it. Once through this layer they travel on to the F layer. This may have split into two as the F1 and F2 layers. When the signals at a frequency of f_3 reach the first of the layers they are again reflected back to earth. Then as the frequency rises to f_4 they pass on to the F2 layer where they are reflected. As the frequency rises still further to f_5 the signals pass straight through all the layers, travelling on into outer space.

During the day at the peak in the sunspot cycle it is possible for signals as high as 50 MHz and more to be reflected by the ionosphere. However, this figure falls to below 20 MHz at very low points in the cycle.

To achieve the longest distances it is best to use the highest layers. This is achieved by using a frequency that is high enough to pass through the lower layers. From this it can be seen that frequencies higher in the short wave spectrum tend to give the longer distance signals. Even so it is still possible for signals to travel from one side of the globe to the other on low frequencies at the right time of day. But for this to happen good antennas are needed at the transmitter and receiver and high powers are generally required at the transmitter.

Distances and the angle of radiation

The distance that a signal travels if it is reflected by the ionosphere is dependent upon a number of factors. One is the height at which it is reflected, and in turn this is dependent upon the layer used for reflection. It is found that the maximum distance for a signal reflected by the E layer is about 2000 km whereas the maximum for a signal reflected by the F layer is about 4000 km.

Signals leave the transmitting antenna at a variety of angles to the earth. This is known as the angle of radiation, and it is defined as the angle between the earth and the path the signal is taking.

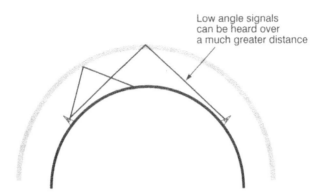

Figure 2.19   *Effect of the angle of radiation on the distance achieved*

It is found that those that have a higher angle of radiation and travel upwards more steeply cover a relatively small distance. Those that leave the antenna almost parallel to the earth travel a much greater distance before they reach the ionosphere, after which they return to the earth almost parallel to the surface. In this way these signals travel a much greater distance.

To illustrate the difference this makes, changing the angle of radiation from 0 degrees to 20 degrees reduces the distance for E layer signals from 2000 km to just 400 km. Similarly using the F layer distances are reduced from 4000 km to 1000 km.

For signals that need to travel the maximum distance this shows that it is imperative to have a low angle of radiation. However, broadcast stations often need to make their antennas directive to ensure the signal reaches the correct area. Not only do they ensure they are radiated with the correct azimuth, they also ensure they have the correct angle of elevation or radiation so that they are beamed to the correct area. This is achieved by altering the antenna parameters.

Multiple reflections

The maximum distance for a signal that is reflected by the F2 layer is about 4000 km. However, radio waves travel much greater distances than this around the world. This cannot be achieved using a single reflection, but instead several are used as shown in Figure 2.20.

To achieve this, the signals travel to the ionosphere and are reflected back to earth in the normal way. Here they can be picked up by a receiver. However, the earth also acts as a reflector because it is conductive and the signals are reflected back to the ionosphere. In fact it is found that areas which are more conductive act as better reflectors. Not surprisingly the

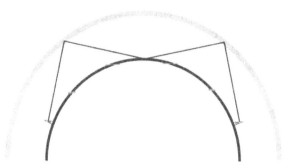

Figure 2.20   *Several reflections used to give greater distances*

sea acts as an excellent reflector. Once reflected at the earth's surface the signals travel towards the ionosphere where they are again reflected back to earth.

At each reflection the signal suffers some attenuation. This means that it is best to use a path that gives the minimum number of reflections as shown in Figure 2.21. Lower frequencies are more likely to use the E layer and as the maximum distance for each reflection is less, it is likely to give lower signal strengths than a higher frequency using the F layer to give fewer reflections.

Not all reflections around the world occur in exactly the ways described. It is possible to calculate the path that would be taken, the number of reflections, and hence the path loss and signal strength expected. Sometimes signal strengths appear higher than would be expected. In conditions like these it is possible that a propagation mode called chordal hop is being experienced. When this happens it is found that the signal travels to the ionosphere where it is reflected, but instead

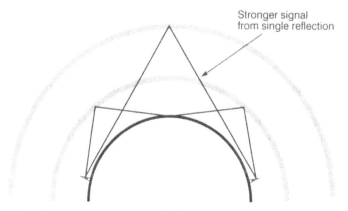

Stronger signal
from single reflection

Figure 2.21   *The minimum number of reflections usually gives the best signal*

of returning to the earth it takes a path which intersects with the ionosphere again, only then being reflected back to earth. Fewer reflections are needed to cover a given distance. As a result signal strengths are higher when this mode of propagation is used.

Critical frequency

When a signal reaches a layer in the ionosphere it undergoes refraction and often it will be reflected back to earth. The steeper the angle at which the signal hits the layer the greater the degree of refraction is required. If a signal is sent directly upwards this is known as vertical incidence as shown in Figure 2.22.

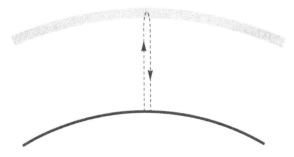

Figure 2.22 *Vertical incidence*

For vertical incidence there is a maximum frequency for which the signals will be returned to earth. This frequency is known as the critical frequency. Any frequencies higher than this will penetrate the layer and pass right through it on to the next layer or into outer space.

MUF

When a signal is transmitted over a long-distance path it penetrates further into the reflecting layer as the frequency increases. Eventually it passes straight through. This means that for a given path there is a maximum frequency that can be used. This is known as the maximum usable frequency or MUF. Generally the MUF is three to five times the critical frequency depending upon which layer is being used and the angle of incidence.

For optimum operation a frequency about 20 per cent below the MUF is normally used. It is also found that the MUF varies greatly depending upon the state of the ionosphere. Accordingly it changes with the time of day, season, position in an 11-year sunspot cycle, and the general state of the ionosphere.

LUF

When the frequency of a signal is reduced, further reflections are required and losses increase. As a result there is a frequency below which the signal cannot be heard. This is known as the lowest usable frequency or LUF.

Skip zone

When a signal travels towards the ionosphere and is reflected back towards the earth, the distance over which it travels is called the skip distance as shown in Figure 2.23. It is also found that there is an area over which the signal cannot be heard. This occurs between the position where the signals start to return to earth and where the ground wave cannot be heard. The area where no signal is heard is called the skip or dead zone.

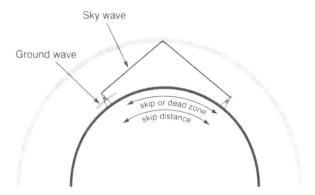

Figure 2.23 *Skip zone and skip distance*

State of the ionosphere

Radio propagation conditions are of great importance to a vast number of users of the short wave bands. Broadcasters, for example, are very interested in them, as are other professional users. To detect the state of the ionosphere an instrument called an ionosonde is used. This is basically a form of radar system that transmits pulses of energy up into the ionosphere. The reflections are then monitored and from them the height of the various layers can be judged. Also, by varying the frequency of the pulses, the critical frequencies of the various layers can be judged.

Fading

One of the characteristics of listening to short wave stations is that some signals appear to fade in and out all the time. These alterations are taken as a matter of course by listeners who are generally very tolerant to the imperfections in the quality of the signal received from the ionosphere. There are a number of different causes for fading but they all result from the fact that the state of the ionosphere is constantly changing.

The most common cause of fading occurs as a result of multipath interference. This occurs because the signal leaves the antenna at a variety of different angles and reaches the ionosphere over a wide area. As the ionosphere is very irregular the signal takes a number of different paths as shown in Figure 2.24. The changes in the ionosphere cause the lengths of these different paths to vary. This means that when the signals come together at the receiving antenna they pass in and out of phase with one another. Sometimes they reinforce one another, and then at other times they cancel each other out. This results in the signal level changing significantly over periods of even a few minutes.

Another reason for signal fading arises out of changes in polarization. It is found that when the ionosphere reflects signals back to earth they can be in any polarization. For the best reception, signals should have the same polarization as the receiving antenna. As the polarization of the reflected wave will change dependent upon the ionosphere, the signal strength will vary according to the variations in polarization.

In some instances the receiver may be on the edge of the skip zone for a particular signal. When this happens any slight variations in the state of

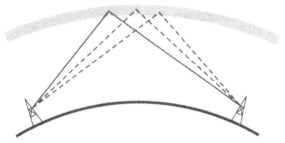

Figure 2.24   *Signals can reach the receiver via several paths*

the ionosphere will cause the receiver to pass into or out of the skip zone, giving rise to signal strength variations.

On other occasions severe distortion can be heard particularly on amplitude modulated signals. This can occur when different sideband frequencies are affected differently by the ionosphere. This is called selective fading and it is often heard most distinctly when signals from the ground and sky waves are heard together.

Ionospheric disturbances

At certain times ionospheric propagation can be disrupted and signals on the short wave bands can completely disappear. These result from disturbances on the sun called solar flares. These flares are more common at times of high sunspot activity, but they can occur at any time.

When a flare occurs there is an increase in the amount of radiation that is emitted. The radiation reaches the earth in about eight minutes and causes what is termed a sudden ionospheric disturbance (SID). This is a fast increase in the level of absorption in the D layer lasting anywhere

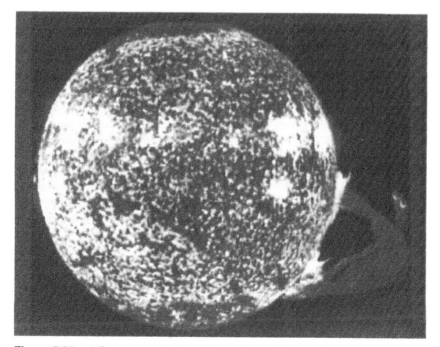

Figure 2.25 *A flare appearing from the surface of the sun (courtesy NASA/Caltech/JPL)*

from a few minutes to a few hours. This can affect all or part of the short wave spectrum, dependent upon the level of increase in radiation.

The next stage of the process sees changes in the solar wind. Under normal conditions there is a flow of particles away from the sun. This is the solar wind, and the earth's magnetic field is able to give protection against this. However, after a flare there is a considerable increase in the solar wind. This normally occurs about 20 to 30 hours after the flare. When it arrives it starts a complicated chain of events. Large variations in the earth's magnetic field can be observed and a visible aurora may be seen in locations towards the poles. Generally it is necessary to be at a latitude of greater than about 55 degrees to see this. Although the short wave bands may initially improve after the SID, the increase in solar wind causes a major degradation in communications over the HF portion of the spectrum. This mainly results from a drastic decrease in the level of ionization in the ionosphere including the D layer that absorbs signals. As a result radio signals are not reflected back to earth in the usual way, causing a radio blackout.

During some stages of the aurora very high levels of ionization are seen towards the poles. As a result signals may be reflected back to earth in these regions at frequencies up to about 150 MHz, although HF communications will be absorbed. When signals are reflected in this way they generally have a distinctive buzz superimposed upon them. This results from the constantly changing nature of the ionosphere under these conditions.

The blackout in HF radio communications may last anywhere from a few hours to a few days after which the bands slowly recover. The first signs of the end of the blackout are normally seen at the low end of the spectrum first. It is also found that further disruption may occur after 28 days, the period of the sun's rotation.

Very low frequency propagation

Propagation of long radio waves is of importance for some long-distance communications and also for some long-distance navigation. In recent years considerable progress has been made in the understanding of the way in which the earth and the ionosphere act as a waveguide at these frequencies.

However, for a more simplified approach the way in which propagation occurs can be considered in a number of ways. For short distances the signal is received mainly as a result of ground wave propagation and it is found that the intensity is virtually inversely proportional to the distance between the transmitter and the receiver. However, beyond a certain point the signal falls at a greater rate because of the earth's curvature and losses in the ground. At large distances the received signal is chiefly due to reflected signals from the ionosphere. As might be expected at intermediate distances the received signal results from a combination of both modes and this results in an interference pattern. At very low frequencies the D layer reflects rather than absorbs signals.

VHF and above

At frequencies above the limit of ionospheric propagation but below about 3000 MHz communication can be established over distances greater than the ordinary line of sight. This is as a result of effects within the troposphere. As most of the conditions that govern our weather occur in the troposphere, there are usually many links between the weather and radio propagation conditions at these frequencies.

Under normal conditions signals at these frequencies travel more than the line of sight. Prior to the 1940s it was generally thought that communication over distances greater than the line of sight was not possible, but experience soon showed this was not true. As a very rough guide it is usually possible to achieve distances at least a third greater than this. This is possible because of the varying refractive index of the air above the earth's surface. An increase in the pressure and humidity levels close to the earth's surface mean that the refractive index of the air is greater than that of the air higher up. Like light waves radio waves can be refracted, and they bend towards the area of the greater refractive index. This means that the signals tend to follow the curvature of the earth and travel over distances that are greater than just the line of sight. An additional effect is that of diffraction where the signal diffracts around the earth's curvature.

Greater distances

At times signals can be received over much greater distances than 4/3 of the line of sight. At times like these terrestrial television channels may be subject to interference as may other radio users. There are a variety of mechanisms by which signals can be propagated over these greater distances. Usually it is possible to predict when there is a likelihood of them occurring as there is a strong correlation between them and certain weather conditions. Usually the extended distances result from the normal gradient in refractive index becoming much steeper. In this way the degree of bending is increased, allowing the signals to follow the curvature of the earth for greater distances.

A number of weather conditions may cause this increase. An area high of pressure may cause the conditions that can increase the normal propagation distance. A high pressure is normally associated with warm weather, especially in summer. Under these conditions the hot air rises and cold air comes in to replace it. This accentuates the density gradient normally present and the change in refractive index occurring as a result of this can be very sharp.

Other weather conditions can also bring about similar increases in the change of refractive index. Cold weather fronts can have the same effect. Here a mass of warm air and a mass of cold air meet. When this occurs the warm air rises above the cold air bringing about similar conditions. Cold weather fronts normally move more quickly than high pressure areas, and as a result increase in propagation distance due to cold fronts is normally more short lived than those caused by high pressure areas.

Other local conditions may give rise to increases in propagation distance. Convection in coastal areas in warm weather, the rapid cooling of the earth and the air closest to it after a hot day, or during frosty weather. Subsidence of cool moist air into valleys on calm summer evenings can give rise to these changes.

Sometimes the changes in refractive index can trap the signals between two layers forming a type of duct or waveguide. When this happens signals may be carried for several hundreds of kilometres.

It is found that tropospheric bending and ducting is experienced more at higher frequencies. Its effects are comparatively small at frequencies at the top of the HF portion of the spectrum, and increases steadily into the VHF and UHF portion of the spectrum. At higher frequencies the effects are still noticed, but other factors start to limit the range.

Troposcatter

The effects of tropospheric bending are very dependent upon the weather. This is shown by the fact that television signals in the UHF band are only occasionally affected by interference from distant signals. As such it is not possible to rely on these modes for extending the range of a communications link. Where links are required a mode of propagation known as troposcatter can be used. This form of propagation relies on the fact that within the troposphere there are masses of air with a slightly different refractive index which are moving around randomly. These arise because of the continually moving nature of air, and the differing temperatures of different parts.

These masses of air reflect and bend the signals and small amounts of the signal are returned to earth. In view of the small amounts of signal which are returned to earth using this mode, high transmitter powers,

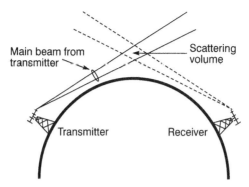

Figure 2.26 *The mechanism behind troposcatter*

high gain antennas and sensitive receivers are required. Nevertheless this form of propagation can be very useful for links over distances around 1000 to 1500 km.

Sporadic E

Sometimes in summer it is possible for signals to be audible in the bands at the top end of the short wave spectrum at the bottom of the sunspot cycle. When the maximum usable frequency may be well below the frequencies in question signals from stations up to 2000 km distance may be heard in summer. This occurs as a result of a form of propagation known as sporadic E.

Sporadic E used to be well known when television transmissions used frequencies around 50 MHz, and sometimes in summer reception would be disturbed by interference from distant stations. Even today reception of VHF FM signals can be disturbed when frequencies around 100 MHz are affected by it. The maximum frequencies it generally affects are up to around 150 MHz, although it has affected higher frequencies just over 200 MHz on very rare occasions.

Sporadic E occurs as a result of highly ionized areas or clouds forming within the E layer. These clouds have a very patchy structure and may measure anywhere between 100 km and 1000 km across and less than a kilometre thick. This means that propagation is quite selective when the clouds are small with signals coming from a particular area. However, their electron density is much greater than that normally found in the E layer and as a result signals with much higher frequencies are reflected. It is also found that there are irregularities in the structure of the clouds and this makes them opaque to lower frequency signals.

At the onset of propagation via sporadic E the level of ionization starts to build up. At first only the less high frequencies are affected. Those at the top end of the HF part of the spectrum are affected first. As the levels of ionization increase further, frequencies into the VHF region are reflected.

In temperate regions, sporadic E normally occurs in the summer, reaching a peak broadly around mid summer. Even so frequencies at the top of the short waveband may be affected on some days at least a couple of months either side of this. Frequencies well up into the VHF portion of the spectrum are normally affected closer to the centre of the season because much high ionization levels are required. It is also found that the very high frequencies are not affected for as long. Sometimes signals may only be heard for a few minutes before propagation is no longer supported.

The sporadic nature of this form of propagation means that it is very difficult to predict when it will occur. Even when propagation is supported by this mode it is very variable. The ionized clouds are blown about in the upper atmosphere by the swiftly moving air currents. This means that the area from which stations are heard can change. Accordingly sporadic E is not a mode normally used for commercial communications.

Meteor scatter

Meteor scatter or meteor burst communication is a useful form of propagation for distances of up to about 2000 km. It is generally used for data links and for applications where real time communications are not required for which it provides a cost effective method of communication.

Meteor scatter relies upon the fact that meteors are constantly entering the earth's atmosphere. It is estimated that about 75 million enter every day. The vast majority of them are small, and do not produce the characteristic visible trail in the sky. In fact most meteors are only about the size of a grain of sand and any that are an inch across are considered to be large.

The meteors enter the atmosphere at speeds of up to 75 km/second and as the atmosphere becomes more dense they burn up, usually at heights of around 80 km. The heat generated from the friction from the air causes the atoms on the surface of the meteor to vaporize. The atoms become ionized and in view of the speeds, leave a trail of positively charged ions and negatively charged atoms.

The trails do not normally last for very long. As the density of the air is relatively high, the electrons and ions are able to recombine quickly. As

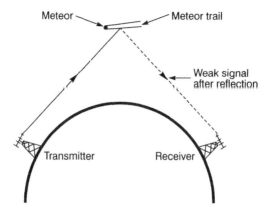

Figure 2.27 *Meteor scatter link*

a result the trails normally only last for a second or so. However, the level of ionization is very high and they are able to reflect radio waves up to frequencies of 100 MHz and more. While the level of ionization is very high, the area which can reflect signals is very small, and only a small amount of energy is reflected. Despite this there is just enough for a sensitive receiver to receive.

The meteors come from the sun, and there are two main types. Most enter the atmosphere on a random basis, while others are from meteor showers. The showers occur at specific times of the year and occur as the earth passes through areas around the sun where there is a large amount of debris.

A wide range of frequencies can support meteor scatter communications, although at lower frequencies signals suffer from attenuation in the D layer of the ionosphere. Also if frequencies in the HF portion of the spectrum are used then there is the possibility of propagation by reflection from the ionosphere. These two reasons mean that meteor scatter operation is generally confined to frequencies above 30 MHz. Operating above these frequencies has the further advantage that both galactic and artificial noise are less – a vital factor when considering the low signal levels involved in this mode of communication.

Generally most meteor scatter operation takes place between 40 and 50 MHz although there is some between 30 and 40 MHz. The top limit is governed more by the fact that television transmissions previously occupied frequencies above 50 MHz, and still do in some countries.

Frequencies above 3 GHz

At frequencies above about 3 GHz, the distances that can be achieved are not normally much in excess of the line of sight. This means that if greater distances are to be achieved, antennas must be placed higher above the earth's surface to increase the distance of the horizon.

Other effects are also noticed. Signals are absorbed more by atmospheric conditions. Rain causes signals at these frequencies to undergo attenuation. The level of attenuation is dependent upon the frequency in use and the rate at which the rain is occurring. Gases also cause signal attenuation. There are peaks in the level of absorption due to water vapour at frequencies of 20 GHz and others at much higher frequencies around 200 and 350 GHz. Similarly oxygen gives rise to peaks in attenuation around 60 GHz and another just over 100 GHz.

3 Modulation

Radio signals are used to carry information. This information may be sound as in broadcasting or mobile telecommunications, or it may be in another form as in a weather fax, or digital data used for sending text messages, for example. In fact there is an almost infinite variety of uses for radio signals, but in all cases they are used to carry information in one form or another.

To be able to transmit any form of information a radio signal or carrier is first generated. The information in the form of audio, data, or other form of signal is used to modify (modulate) the carrier and in this way the information is superimposed onto the carrier and is transmitted to the receiver. Here the information is removed from the radio signal and reconstituted in its original format in a process known as demodulation. It is worth noting at this stage that the carrier itself does not convey any information.

There are many forms of modulation that can be used to modify a carrier. Each one has its own advantages and disadvantages and can perform well under given circumstances. Some of the simpler forms of transmission have the advantage that receivers needed to resolve them properly are not as complicated. On the other hand, other modes which need more complicated circuitry to resolve them may perform better in one aspect or another.

Radio carrier

The basis of any radio signal or transmission is the carrier. This consists of an alternating waveform like that shown in Figure 3.1. This is generated in the transmitter and is usually passed to the output of the transmitter where it is radiated for reception by the receiver. The transmission may be anywhere within the radio spectrum described in Chapter 2.

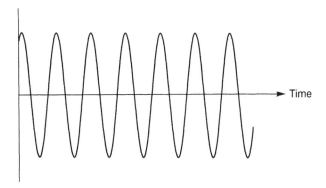

Figure 3.1 *An alternating waveform*

Morse

Morse is the oldest and simplest way of transmitting information using radio. Yet despite its age it still has some advantages, and this means that it is still in use in some areas today.

One of its advantages is its simplicity. It only consists of a carrier wave which is turned on and off as shown in Figure 3.2. The characteristic dots and dashes are defined by the length of time the transmission is left on. The dots and dashes then go to make up the required letters.

The simplicity of the Morse code and its implementation means that equipment for sending it can be much simpler than if another mode had

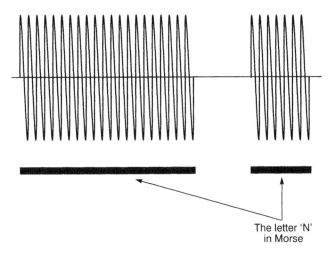

The letter 'N'
in Morse

Figure 3.2 *A Morse signal*

been used. This can be an advantage for amateur radio enthusiasts as it enables them to build their equipment more easily.

Morse has a number of technical advantages. Its relatively slow signalling rate means that it occupies a narrow bandwidth. As a result very narrow filters can be used to cut out most of the interference. It is also found that because the human brain only has to detect the presence or absence of a signal, it can be read at a lower level than a signal carrying speech. These two factors together mean that Morse can be copied at much lower signal levels than several other forms of transmission.

If Morse is received by an ordinary domestic portable radio receiver it is simply heard as a series of clicks and pops as the signal is turned on and off. To make the characteristic tone, a circuit called a beat frequency oscillator (BFO) is needed. This circuit generates a signal that mixes or beats with the incoming signal to make the characteristic Morse sounds of a Morse signal.

These days the use of Morse is relatively restricted. It is no longer used as a primary mode for maritime purposes, as most ships these days use satellite communications systems that use other modes of transmission. Other services including the military have also ceased to use it, and as such its use is confined to a very few professional applications and radio amateurs.

Amplitude modulation

Morse code has its advantages but we have all become accustomed to hearing music and speech over the radio. There are a number of ways in which a carrier can be modulated to take an audio signal. The most obvious way is to change its amplitude in line with the variations in intensity of the sound wave. In this way the overall amplitude or envelope of the carrier is modulated to carry the audio signal as shown in Figure 3.3. Here the envelope of the carrier can be seen to change in line with the modulating signal.

Amplitude modulation or AM is one of the most straightforward methods of modulating a signal. Demodulation, or the process where the radio frequency signal is converted into an audio frequency signal, is also very simple. It only requires a simple diode detector circuit like that shown in Figure 3.4. In this circuit the diode rectifies the signal, only allowing the one half of the alternating radio frequency waveform through. A capacitor is used to remove the radio frequency parts of the signal, leaving the audio waveform. This can be fed into an amplifier after which it can be used to drive a loudspeaker. As the circuit used for demodulating AM is very cheap, it enables the cost of receivers for AM to be kept low.

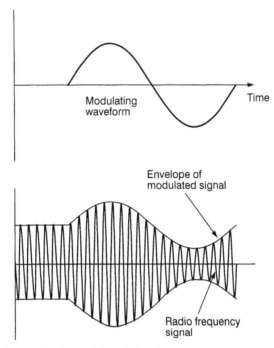

Figure 3.3   *An amplitude modulated signal*

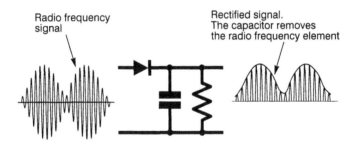

Figure 3.4   *A simple diode detector circuit*

AM has advantages of simplicity, but it is not the most efficient mode to use, both in terms of the amount of spectrum it takes up and the usage of the power. It is for this reason that it is rarely used for communications purposes. Its only major communications use is for VHF aircraft communications. However, it is still widely used on the long, medium, and short wave bands for broadcasting because of its simplicity, enabling the cost of radio receivers to be kept to a minimum. Many radio stations can be heard using AM. For example, within the UK the BBC broadcasts many of its domestic networks on the long and medium wave bands. BBC Radio 4 broadcasts on 198 kilohertz, and then on the short wave bands there are stations including the BBC World Service, Voice of America and many more that use AM.

To find out why it is inefficient it is necessary to look at a little theory behind the operation of AM. When a radio frequency signal is modulated by an audio signal the envelope will vary. The level of modulation can be increased to a level where the envelope falls to zero and then rises to twice the unmodulated level. Any increase on this will cause distortion because the envelope cannot fall below zero. As this is the maximum amount of modulation possible it is called 100 per cent modulation.

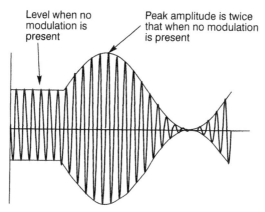

Figure 3.5 *Fully modulated signal*

Even with 100 per cent modulation the utilization of power is very poor. When the carrier is modulated sidebands appear at either side of the carrier in its frequency spectrum. Each sideband contains the information about the audio modulation. To look at how the signal is made up and the relative powers take the simplified case where the 1 kHz tone is modulating the carrier. In this case two signals will be found 1 kHz either side of the main carrier as shown in Figure 3.6. When the carrier is fully modulated, i.e. 100 per cent, the amplitude of the modulation is equal to half that of the main carrier, i.e. the sum of the powers of the sidebands is equal to half that of the carrier. This means that each sideband is just a quarter of the total power. In other words for a transmitter with a 100 watt carrier, the total sideband power would be 50 watts and each individual sideband would be 25 watts. During the modulation process the carrier power remains constant. It is only needed as a reference during the demodulation process. This means that the sideband power is the useful section of the signal, and this corresponds

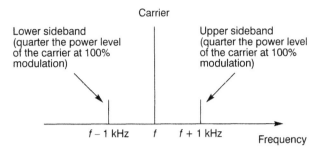

Figure 3.6 *Spectrum of a signal modulated with a 1 kHz tone*

to $(50/150) \times 100$ per cent, or only 33 per cent of the total power transmitted.

Not only is AM wasteful in terms of power, it is also not very efficient in its use of spectrum. If the 1 kHz tone is replaced by a typical audio signal made up of a variety of sounds with different frequencies then each frequency will be present in each sideband. Accordingly the sidebands spread out either side of the carrier as shown and the total bandwidth used is equal to twice the top frequency that is transmitted. In the crowded conditions found on many of the short wave bands today, this is a waste of space, and other modes of transmission which take up less space are often used.

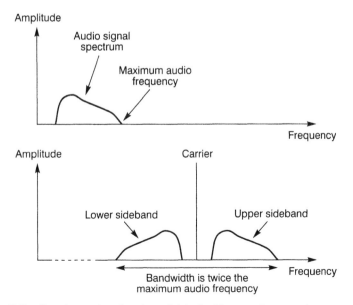

Figure 3.7 *Spectrum of a signal modulated with speech or music*

Modulation index

It is often necessary to define the level of modulation that is applied to a signal. A factor or index known as the modulation index is used for this. When expressed as a percentage it is the same as the depth of modulation. In other words it can be expressed as:

$$M = \frac{\text{RMS value of modulating signal}}{\text{RMS value of unmodulated signal}}$$

The value of the modulation index must not be allowed to exceed one (i.e. 100 per cent in terms of the depth of modulation) otherwise the envelope becomes distorted and the signal will 'splatter' either side of the wanted channel, causing interference and annoyance to other users.

Single sideband

One of the modes widely used for communications traffic is called single sideband. This is a derivative of AM, but by manipulating the signal in the transmitter, the disadvantages of AM can be removed to give a highly efficient mode of transmission.

There are two main stages in the generation of a single sideband signal. The first is that the carrier is removed. This does not contribute to carrying the sound information, and is only used during the demodulation process. As a result it is possible to remove it in the transmitter as shown in Figure 3.8, enabling power to be saved.

It is also found that only one sideband is needed. Both the upper and the lower sidebands are exact mirror images of one another, and either can be used equally well for conveying the sound information. If two sidebands are present on a signal twice the bandwidth is used. To save on the use of bandwidth it is possible to remove one sideband without

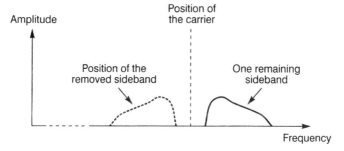

Figure 3.8 *Spectrum of a single sideband signal*

degrading the signal. This can be done because both sidebands carry exactly the same information; they are exact mirror images of one another. The other advantage of transmitting only one sideband is that filter bandwidths can be reduced in the receiver to cut out more unwanted interference and give better reception. It is then possible to remove the carrier to conserve power. This form of modulation is known as single sideband suppressed carrier (SSBSC), or single sideband (SSB) for short.

To demodulate the signal the carrier has to be reintroduced in the receiver using a beat frequency oscillator (BFO) which may also be called a carrier insertion oscillator (CIO). The BFO must be on the correct frequency relative to the sideband being received. Any deviation from this will cause the pitch of the recovered audio to change. While errors of up to about 100 Hz are acceptable for many communications applications, if music is to be transmitted the carrier must be reintroduced on exactly the correct frequency. This can be accomplished by transmitting a small amount of carrier, and using circuitry in the receiver to lock onto this. This is known as single sideband reduced carrier.

As either sideband can be used equally well, a convention is needed for which sideband to use. In this way receivers can be set up to expect the received sideband. If this is not done then the receiver has to be continually switched between upper and lower sideband. For commercial operations the upper sideband is adopted as standard. Although it is perfectly possible to use SSB for frequencies above 30 MHz it is infrequently used on these frequencies, although it is widely used on frequencies below 30 MHz. Signal propagation and the way the bands are used mean that other modes are often more suitable.

It is often necessary to define the output power of a single sideband transmitter. As the output is continually varying and dependent upon the level of modulation at any instant, a measure known as the peak envelope power (PEP) is used. This is the peak level of power of the transmitted signal and includes the sideband plus any pilot carrier that may be included.

Frequency modulation

The most obvious method of applying modulation to a signal is to superimpose the audio signal onto the amplitude of the carrier. However, this is by no means the only method which can be employed. It is also possible to vary the frequency of the signal to give frequency modulation or FM. It can be seen from Figure 3.9 that the frequency of the signal varies as the voltage of the modulating signal changes.

The amount by which the signal frequency varies is very important. This is known as the deviation and is normally quoted as the number of

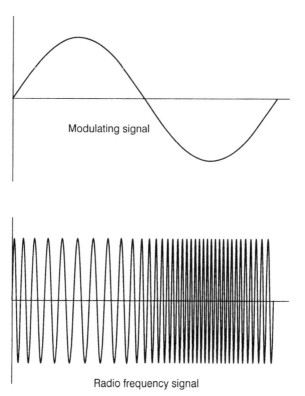

Modulating signal

Radio frequency signal

Figure 3.9 *A frequency modulated signal*

kilohertz deviation. As an example the signal may have a deviation of ±3 kHz. In this case the carrier is made to move up and down by 3 kHz.

Broadcast stations in the VHF portion of the frequency spectrum between 88.5 and 108 MHz use large values of deviation, typically ±75 kHz. This is known as wideband FM (WBFM). These signals are capable of supporting high quality transmissions, but occupy a large amount of bandwidth. Usually 200 kHz is allowed for each wideband FM transmission. For communications purposes less bandwidth is used. Narrow-band FM (NBFM) often uses deviation figures of around ±3 kHz or possibly slightly more.

FM is used for a number of reasons. One particular advantage is its resilience to signal level variations. The modulation is carried only as variations in frequency. This means that any signal level variations will not affect the audio output, provided that the signal does not fall to a level where the receiver cannot cope. As a result this makes FM ideal for mobile or portable applications where signal levels are likely to vary considerably. The other advantage of FM is its resilience against noise and interference. It is for this reason that FM is used for high quality broadcast transmissions.

To demodulate an FM signal it is necessary to convert the frequency variations into voltage variations. This is slightly more complicated than demodulating AM, but it is still relatively simple to achieve. Rather than just detecting the amplitude level using a diode, a tuned circuit has to be incorporated so that a different output voltage level is given as the signal changes its frequency.

Modulation index and deviation ratio

In just the same way that it is useful to know the modulation index of an amplitude modulated signal the same is true for a frequency modulated signal. The modulation index is equal to the ratio of the frequency deviation to the modulating frequency. The modulation index will vary according to the frequency that is modulating the transmitted carrier and the amount of deviation. However, when designing a system it is important to know the maximum permissible values. This is given by the deviation ratio and is obtained by inserting the maximum values into the formula for the modulation index.

$$D = \frac{\text{Max deviation frequency}}{\text{Max modulation frequency}}$$

For a VHF FM sound broadcast transmitter the maximum deviation is 75 kHz and the maximum modulation frequency is 15 kHz giving a deviation ratio of 5.

Sidebands

Any signal that is modulated produces sidebands. In the case of an amplitude modulated signal they are easy to determine, but for frequency modulation the situation is not quite as straightforward. For small values of modulation index, when using narrow-band FM, an FM signal consists of the carrier and the two sidebands spaced at the modulation frequency either side of the carrier. This looks to be the same as an AM signal, but the difference is that the lower sideband is out of phase by 180 degrees.

As the modulation index increases it is found that other sidebands at twice the modulation frequency start to appear. As the index is increased further other sidebands can also be seen. It is also found that the relative levels of these sidebands change, some rising in level and others falling as the modulation index varies. This makes prediction of the exact levels of all the sidebands more difficult than for AM. If a mathematical calculation of the levels of the sidebands is required, a function known as a Bessel function or series needs to be calculated.

Bandwidth

In the case of an amplitude modulated signal the bandwidth required is twice the maximum frequency of the modulation. While the same is true for a narrow-band FM signal, the situation is not true for a wide-band FM signal. Here the required bandwidth can be very much larger, with detectable sidebands spreading out over large amounts of the frequency spectrum. Usually it is necessary to limit the bandwidth of a signal so that it does not unduly interfere with stations either side.

While it is possible to limit the bandwidth of an FM signal, this should not introduce any undue distortion. To achieve this it is normally necessary to allow a bandwidth equal to twice the maximum frequency of deviation plus the maximum modulation frequency. In other words for a VHF FM broadcast station this must be (2 × 75) + 15 kHz, i.e. 175 kHz. In view of this a total of 200 kHz is usually allowed, enabling stations to have a small guard band and their centre frequencies on integral numbers of 100 kHz.

Improvement in signal to noise ratio

It has already been mentioned that FM can give a better signal to noise ratio than AM when wide bandwidths are used. The amplitude noise can be removed by limiting the signal to remove it. In fact the greater the deviation the better the noise performance. When comparing an AM signal to an FM one an improvement equal to $3D^2$ is obtained where D is the deviation ratio.

Pre-emphasis and de-emphasis

An additional improvement in signal to noise ratio can be achieved if the audio signal is pre-emphasized. To achieve this the lower level high frequency sounds are amplified to a greater degree than the lower frequency sounds before they are transmitted. Once at the receiver the signals are passed through a network with the opposite effect to restore a flat frequency response.

To achieve the pre-emphasis the signal is passed through a capacitor–resistor (*CR*) network. At frequencies above the cutoff frequency the signal increases in level by 6 dB per octave. Similarly at the receiver the response falls by the same amount.

Both the receiver and transmitter networks must match one another. In the UK the *CR* time constant is chosen to be 50 μs. For this the break frequency f_1 is 3183 Hz. For broadcasting in North America values of 75 μs with a break frequency of 2.1 kHz are used.

Pre-emphasizing the audio for an FM signal is effective because the noise output from an FM system is proportional to the audio frequency. In order to reduce the level of this effect, the audio amplifier in the receiver must have a response that falls with frequency. In order to prevent the audio signal from losing the higher frequencies, the transmitter must increase the level of the higher frequencies to compensate. This can be achieved because the level of the high frequency sounds is usually less than those lower in frequency.

Frequency shift keying

Many signals heard on the bands will employ a system called frequency shift keying or FSK to carry digital data. Here the frequency of the signal is changed from one frequency to another, one frequency counting as the digital one (mark) and the other as a digital zero (space). By changing the frequency of the signal between these two frequencies it is possible to send data over the radio.

FSK is widely used on the HF bands. To generate the audio tone required from the receiver, a beat frequency oscillator must be used. Accordingly to obtain the correct audio ones the receiver must be tuned to the correct frequency.

Often at frequencies in the VHF and UHF portion of the spectrum a slightly different approach is adopted as frequency stability or accuracy may be a problem. An audio tone is used to modulate the carrier and the audio is shifted between the two frequencies. Although the carrier can be amplitude modulated, frequency modulation is virtually standard. Using audio frequency shift keying (AFSK) the tuning of the receiver becomes less critical.

When the data signal leaves the receiver it is generally in the form of an audio signal switching between two tones. This needs to be converted into the two digital signal levels. This is achieved by a unit called a modem. This stands for MOdulator/DEModulator. Audio tones fed into the modem from the receiver generate the digital levels required for a computer or other equipment to convert into legible text. Conversely it is able to convert the digital signals into the audio tones required to modulate a transmitter to transmit data.

The speed of the transmission is important. For the receiver to be able to decode the signal it must be expecting data at the same rate it is arriving. Accordingly a number of standard speeds are used. The unit used for this is the baud. One baud is equal to one bit per second.

Codes are used to enable the series of marks and spaces or ones and zeros to be represented as alphanumerics. Early teleprinter systems used a code called the Murray code. Often ASCII (American Standard Code for Information Interchange) is used for the basis of some systems now. Also as FSK systems are widely used on the HF portion of the spectrum where levels of interference are high, the systems use high levels of error checking and acknowledgements when a packet of data has been received. These systems enable the data to be received with very few errors even when some interference is present.

Phase modulation

Another form of modulation that is widely used is called phase modulation. As phase and frequency are inextricably linked (frequency being the rate of change of phase) both forms of modulation are often referred to by the common name of angle modulation.

To explain how phase modulation works it is first necessary to give an explanation of phase. A radio signal consists of an oscillating carrier in the form of a sine wave as shown in Figure 3.10. The amplitude follows this curve moving positive and then negative, returning to the start point after one complete cycle. This can also be represented by the movement of a point around a circle, the phase at any given point being the angle between the start point and the point on the waveform as shown.

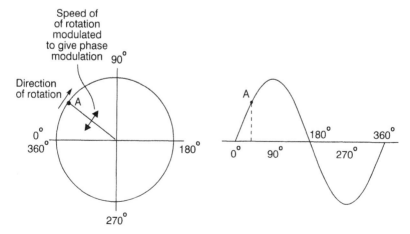

Figure 3.10 *Phase modulation*

Modulating the phase of the signal changes the phase of the signal from what it would have been if no modulation was applied. In other words the speed of rotation around the circle is modulated about the mean value as shown in Figure 3.10. To achieve this it is necessary to change the frequency of the signal for a short time. In other words when phase modulation is applied to a signal there are frequency changes and vice versa. Phase and frequency are inseparably linked as phase is the integral of frequency. Frequency modulation can be changed to phase modulation by simply adding a *CR* network to the modulating signal that integrates the modulating signal.

Phase shift keying

Phase modulation may be used for the transmission of data. Frequency shift keying is robust and has no ambiguities because one tone is higher than the other. However, phase shift keying (PSK) has many advantages in terms of efficient use of bandwidth.

The problem with phase shift keying is that the receiver cannot know the exact phase of the transmitted signal to determine whether it is in a mark or space condition. This would not be possible even if the transmitter and receiver clocks were accurately linked because the path length would determine the exact phase of the received signal. To overcome this problem PSK systems use a differential method for encoding the data onto the carrier. This is accomplished, for example, by making a change in phase equal to a one, and no phase change equal to a zero. Further improvements can be made upon this basic system and a number of other types of phase shift keying have been developed. One simple improvement can be made by making a change in phase by 90 degrees in one direction for a one, and 90 degrees the other way for a zero. This retains the 180 degree phase reversal between one and zero states, but gives a distinct change for a zero. In a basic system not using this process it may be possible to lose synchronization if a long series of zeros are sent. This is because the phase will not change state for this occurrence.

There are many variations on the basic idea of phase shift keying. Each one has its own advantages and disadvantages enabling system designers to choose the one most applicable for any given circumstances.

Minimum shift keying

It is found that binary data consisting of sharp transitions between 'one' and 'zero' states and vice versa potentially creates signals that have sidebands extending out a long way from the carrier, and this is not ideal from many aspects. This can be overcome in part by filtering the signal, but is found that the transitions in the data become progressively less sharp as the level of filtering is increased and the bandwidth reduced. To overcome this a form of modulation known as Gaussian filtered minimum shift keying (GMSK) is widely used, and has, for example, been adopted for the GSM standard for mobile telecommunications. It is derived from a modulation scheme known as minimum shift keying (MSK) which is what is known as a continuous phase scheme. Here there are no phase discontinuities because the frequency changes occur at the carrier zero crossing points.

To illustrate this take the example shown in Figure 3.11. Here it can be seen that the modulating data signal changes the frequency of the signal and there are no phase discontinuities. This arises as a result of the unique factor of MSK that the frequency difference between the logical one and logical zero states is always equal to half the data rate. This can be expressed in terms of the modulation index, and it is always equal to 0.5.

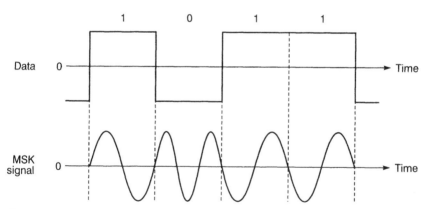

Figure 3.11   *An example of an MSK signal*

While this method appears to be fine, it is found that the bandwidth occupied by an MSK signal is too wide for many systems where a maximum bandwidth equal to the data rate is required. A plot of the spectrum of an MSK signal shows sidebands extending well beyond a bandwidth equal to the data rate. This can be reduced by passing the modulating signal through a low-pass filter prior to applying it to the carrier. The requirements for the filter are that it should have a sharp cutoff, narrow bandwidth and its impulse response should show no overshoot. The ideal filter is known as a Gaussian filter which has a Gaussian-shaped response to an impulse and no ringing.

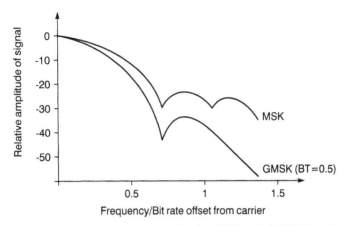

Figure 3.12   *Graph of the spectral density for MSK and GMSK signals*

There are two main ways in which GMSK can be generated. The most obvious is to apply the digital modulating signal filtered by a Gaussian filter and then applied to a frequency modulator where the modulation index is set to 0.5 as shown in Figure 3.13. While simple, this method has the drawback that the modulation index must exactly equal 0.5. In practice this analogue method is not suitable because component tolerances drift and cannot be set exactly.

A second method is more widely used. Here what is known as a quadrature modulator is used. The term quadrature means that the phase of a signal is in quadrature or 90 degrees to another one. The quadrature modulator uses one signal that is said to be in-phase and another that is in quadrature to this. In view of the in-phase and quadrature elements this type of modulator is often said to be an I-Q modulator. Using this type of modulator the modulation index can be maintained at exactly 0.5 without the need for any settings or adjustments. This makes it much easier to use, and capable of providing the required level of performance without the need for adjustments. For demodulation the technique can be used in reverse.

A further advantage of GMSK is that it can be amplified by a non-linear amplifier and remain undistorted. This is because there are no elements

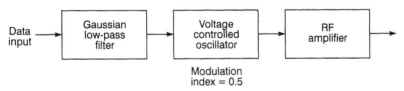

Figure 3.13   *Generating GMSK using a Gaussian filter and a frequency modulator with the modulation index set to 0.5*

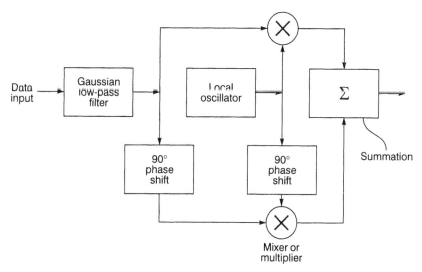

Figure 3.14 *A block diagram of a quadrature or I-Q modulator used to generate GMSK*

of the signal that are carried as amplitude variations. Furthermore the system is capable of providing a low level of bit errors, i.e. a low bit error rate (BER) under noisy conditions.

Pulse modulation

While amplitude and angle modulation, together with their many derivatives, form the vast majority of transmissions, another type of modulation, known as pulse modulation is used in many instances. The advantage of this type of modulation is that a number of different signals can be transmitted together using a system known as time division multiplexing.

Using this type of modulation, short periodic samples are transmitted. The modulating waveform is sampled as shown in Figure 3.15(a) and this information is used to change the characteristics of the pulses in one of a variety of ways. Once at the receiver the samples are used to reconstitute the original signal which can be used as required. Surprisingly few samples are needed and it can be shown that if an audio signal is sampled at a frequency just over twice the highest frequency present, then the samples will contain all the necessary information to recreate the signal in its original format. Take as an example the waveform shown in Figure 3.15(b). If the waveform was sampled once a cycle then each sample would have the same value and there would be no indication about the

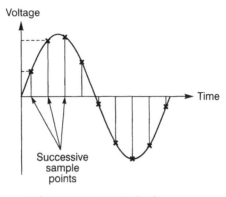

(a) Samples taken periodically

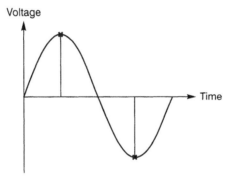

(b) Minimum number of samples

Figure 3.15   *Sampling a waveform*

actual waveform. However, if the waveform is sampled at twice the frequency of the waveform, the samples alternate either side of the zero axis, and it can be deduced that it is varying and has a certain frequency.

There are a number of ways in which pulses derived from samples of a waveform can be used to modulate a carrier. Possibly the most obvious is pulse amplitude modulation (PAM) where the amplitude of the pulses represents the amplitude of the waveform at that point as shown in Figure 3.16. It can also be seen that it is possible to place more than one set of modulating pulses on the signal, by spacing them apart in time. This concept is known as time division multiplexing. At the transmitting end of the link the pulses are placed onto the carrier in a given order. By taking the pulses of carrier at the receiver end in the same order, each signal can be properly reconstituted.

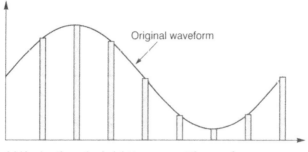

(a) Varying the pulse height to represent the waveform

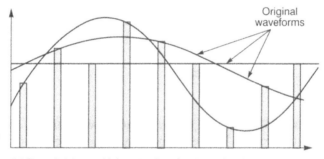

(b) Time division multiplexed pulses for three signals

Figure 3.16 *Pulse amplitude modulation*

Another method of pulse modulation involves varying the width of the pulses according to the amplitude of the waveform. This is called pulse width modulation (PWM). An example is shown in Figure 3.17 and from this it can be seen that the width of the pulses increases in line with the instantaneous value of the waveform.

Both PAM and PWM are essentially analogue methods of pulse modulating a signal. There are many advantages to digitizing the signal and using a digital train of pulses to modulate the carrier. This form of pulse code modulation (PCM) is widely used for telecommunications links. The advantage is that once the signal is digitized, the system only has to recognize whether a pulse is present or not, i.e. a logical one or a zero. If the noise level is not large then there is a very small chance of there being any errors. It is found that if the RMS noise level is 20 dB below the peak pulse level then there is a negligible chance of there being an error and at 13 dB the chance of an error becomes one in a hundred. As a result PCM systems are more resilient to noise than their analogue equivalents. This has the advantage that repeater stations to boost the signals are required less frequently giving a significant saving in costs.

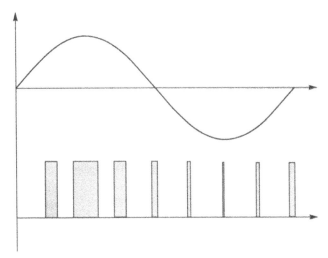

Figure 3.17   *Pulse width modulation*

To generate a PCM signal, the modulating waveform is sampled as before at distinct points. These samples are then converted into a digital format using a circuit called an analogue to digital converter. This circuit generates a binary code that in turn is converted into a group of pulses. A group of n pulses can represent 2^n different levels, i.e. if three pulses are generated then there can be 2^3 or 8 levels. In reality eight pulses are usually used for communications links giving a total of 256 levels.

To reconstitute the waveform the incoming pulses are converted into a binary code that enters a circuit called a digital to analogue converter. This regenerates the audio signal into a form of staircase waveform with steps as shown. The steps can be removed by adding a capacitor circuit to smooth the waveform, giving a very close approximation to the original signal.

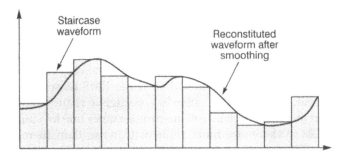

Figure 3.18   *Regenerating an analogue signal from its digital representation*

The main disadvantage of PCM when compared to PAM or PWM is that far more pulses are transmitted, and this means that a much greater bandwidth is required. If eight pulses are used to represent each sample then the bandwidth will be eight times as much. However, the increase in noise immunity it provides usually more than outweighs the problems with the bandwidth.

Spread spectrum techniques

In many instances it is necessary to keep transmissions as narrow as possible to conserve the frequency spectrum. However, under some circumstances it is advantageous to use what are known as spread spectrum techniques where the transmission is spread over a wide bandwidth. There are two ways of achieving this. One is to use a technique known as frequency hopping, while the other involves spreading the spectrum over a wide band of frequencies so it appears as background noise. These can be done in different ways and the two most widely used systems are DSSS and OFDM.

Frequency hopping

In some instances, particularly in military applications, it is necessary to prevent any listeners apart from those intended from picking up a signal or from jamming it. It may also be used to reduce levels of interference. If interference is present on one channel, the hopping signal will only remain there for a short time and the effects of the interference will be short lived. Frequency hopping is a well-established principle. In this system the signal is changed many times a second in a pseudo-random sequence from a predefined block of channels. Hop rates vary dependent upon the requirements. Typically the transmission may hop a hundred times a second, although at HF this will be much less.

The transmitter will remain on each frequency for a given amount of time before moving on to the next. There is a small dead time between the signal appearing on the next channel, during this time the transmitter output is muted. This is to enable the frequency synthesizer time to settle, and to prevent interference to other channels as the signal moves.

To receive the signal the receiver must be able to follow the hop sequence of the transmitter. To achieve this both transmitter and receiver must know the hop sequence, and the hopping of both transmitter and receiver must be synchronized.

Frequency hopping transmissions usually use a form of digital transmission. Even when speech is used, this has to be digitized before being sent. The data rate over the air has to be greater than the overall throughput. This is to allow for the dead time while the set is changing frequency.

Direct sequence spread spectrum

Direct sequence spread spectrum (DSSS) is a form of spread spectrum modulation that is being used increasingly as it offers improvements over other systems, although this comes at the cost of greater complexity in the receiver and transmitter. It is used for some military applications where it provides greater levels of security and it has been chosen for many of the new cellular telecommunications systems where it can provide an improvement in capacity. In this application it is known as code division multiple access because it is a system whereby a number of different users can gain access to a receiver as a result of a different 'code' they use. Other systems use different frequencies (frequency division muptiple access – FDMA) or different times or time slots on a transmission (time division multiple access – TDMA).

Its operation is more complicated than those that have already been described. When selecting the required signal there has to be a means by which the selection occurs. For signals such as AM and FM, different frequencies are used, and the receiver can be set to a given frequency to select the required signal. Other systems use differences in time. For example, using pulse code modulation, pulses from different signals are interleaved in time, and by synchronizing the receiver and transmitter to look at the overall signal at a given time, the required signal can be selected. CDMA uses different codes to distinguish between one signal and another. To illustrate this, take the analogy of a room full of people speaking different languages. Although there is a large level of noise, it is possible to pick out the person speaking English, even when there may be people who are just as loud, or may be even louder speaking a different language you may not be able to understand.

The system enables several sets of data to be placed onto a carrier and transmitted from one base station, as in the case of a cellular tele-communications base station. It also allows for individual units to send data to a receiver that can receive one of more of the signals in the presence of a large number of others.

To achieve this the signal is spread over a given bandwidth. This is achieved by using a spreading code. This operates at a higher rate than the data. The codes for this can either be random, of they can be what is known as orthogonal. Orthogonal codes are ones which when multiplied together and then added up over a period of time have a sum of zero. To illustrate this, take the example of two codes:

Code A	1	–1	–1	1
Code B	1	–1	1	–1
Product	1	1	–1	–1

summed over a period of time = 0, i.e. $1 + 1 - 1 - 1 = 0$

Using orthogonal codes it is possible to transmit a large number of data channels on the same signal. To achieve this the data is multiplied with the chip stream. This chip stream consists of the codes being sent repeatedly so that each data bit is multiplied with the complete code in the chip stream. In other words if the chip stream code consists of four bits then each data bit will be successively multiplied by four chip bits. It is also worth noting that the spread rate is the number of data bits in the chip code, i.e. the number of bits that each data bit is multiplied by. In this example the spread rate is four because there are four bits in the chip code.

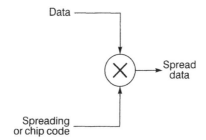

Figure 3.19 *Multiplying the data stream with the chip stream*

To produce the final signal that carries several data streams the outputs from the individual multiplication processes are summed. This signal is then converted up to the transmission frequency and transmitted.

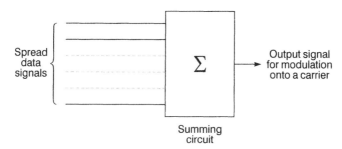

Figure 3.20 *Generating a signal that carries several sets of data*

At the receiver the reverse process is adopted. The signal is converted down to the base-band frequency. Here the signal is multiplied by the relevant chip code to extract the relevant data in a process known as correlation. By multiplying by a different chip code a different set of data will be extracted.

To illustrate how this operates it is best to provide an example showing how the signal is generated, and then the original data extracted from the received signal.

Chip code A   1 -1 -1   1

Chip code B   1 -1   1 -1

| Data stream 1 | 1 | | | -1 | | | 1 | | | 1 | | |

This makes
chip stream: 1 -1 -1 1 |-1 1 1 -1 | 1 -1 -1 1 | 1 -1 -1 1 |

Data stream 2   -1 | 1 | 1 | 1

This makes the
chip stream: -1 1 -1 1 | 1 -1 1 -1 | 1 -1 1 -1 | 1 -1 1 -1 |

Adding the two chip streams the sum that is transmitted becomes:

0   0 -2   2 | 0   0   2 -2 | 2 -2   0   0 | 2 -2   0   0 |

In the receiver this needs to be multiplied by the relevant code. For the example take code A:

0   0 -2   2 | 0   0   2 -2 | 2 -2   0   0 | 2 -2   0   0 |

multiply by   1 -1 -1   1 | 1 -1 -1   1 | 1 -1 -1   1 | 1 -1 -1   1 |

and this becomes 0   0   2   2 | 0   0 -2 -2 | 2   2   0   0 | 2   2   0   0 |

add the sum of each group

4 | -4 | 4 | 4

This equates to the original signal of

1 | -1 | 1 | 1

When random or more correctly a pseudo-random spreading code is used a similar process is followed. Instead of using the orthogonal codes, a pseudo-random spreading sequence is used. Both the transmitter and receiver will need to be able to generate the same pseudo-random code.

This is easily achieved by ensuring that both transmitter and receiver use the same algorithms to generate these sequences. The drawback of using a pseudo-random code is that the codes are not orthogonal and as a result some data errors are expected when regenerating the original data.

OFDM

Another form of modulation that is becoming more frequently used with the rise in levels of integration is called orthogonal frequency division muliplex (OFDM). A form of this known as coded OFDM or COFDM is used for digital radio (DAB) broadcasts that are being made and is applicable for use in the cellular telecommunications environment. COFDM has forward error correction applied to the signal before transmission and this enables it to overcome errors caused by interference and lost carriers due to selective fading or reflections to be overcome. As a result of its resilience to interference and reflected signals that introduce data delays, it is an ideal medium for broadcasting digital radio. It caters equally for those with hi-fi tuners at home and those in motor vehicles that are on the move and require omnidirectional antennas. These mobile antennas can pick up the signals that may be arriving from a variety of directions due to reflections, and hence may not all arrive at the same time.

OFDM is a form of transmission that uses multiple carriers. The data to be transmitted is spread across the carriers so that each one carries a low data rate stream. The carriers are spaced very closely to preserve the bandwidth. Normally they would interfere with one another but the signals are made orthogonal to each other to prevent this occurring. This is achieved by making the data bit rate equal to the carrier spacing. In this way the nulls in the sidebands occur at the point where the next signal occurs. This arises because the spectrum of the signal from each carrier has a sin x/x shape, and the next carrier is slotted in where the null in the spectrum of the first carrier exists.

Summary

There are two basic ways in which a signal can be modulated. Either its amplitude or phase/frequency can be varied. However, there are a great many ways in which this can be achieved, each type has its advantages and disadvantages. Accordingly the choice of the correct type of modulation is critical when designing a new system.

4 Antenna systems

Every receiver or transmitter requires an antenna if it is to receive or radiate radio signals. In some instances these antennas may be small, and incorporated into the equipment making them almost invisible as in the case of mobile phones. In other instances large antennas are used. Some good examples of very large antenna systems are used by short wave broadcast stations. Others might include a satellite communications centre such as Goonhilly in Cornwall, UK, the antennas used by NASA for communications with various space missions like that located at the Goldstone Deep Space Communications Complex or possibly a radio astronomy antenna. However, a vast number of antennas are more modest in size and are used in many applications. One very common example is a television antenna that can be seen on many domestic houses around the world. Despite the size of an antenna, its performance is of paramount importance. Whether small or large, every antenna needs to be optimized for its particular application to ensure that its performance is maximized.

The function of the antenna for a receiver is to pick up the radio electromagnetic waves and convert them into electrical signals. Once they exist as electrical signals they are transferred from the antenna element itself into the receiver where they are amplified, filtered and demodulated to give the required audio output. Conversely in terms of a transmitter the antenna performs the function of converting the electrical energy into radiated electromagnetic signals.

Antennas have a variety of properties. They can only operate efficiently over a given bandwidth. They also have an electrical impedance, and they are polarized, only picking up waves of a certain polarization. Many of these factors are very important and if the antenna is to give the optimum performance it is necessary to ensure it has the correct properties and that it is set up correctly.

Resonance and bandwidth

An antenna is a form of tuned circuit that acts like an inductor at some frequencies and a capacitor at others. Like a tuned circuit it has a resonant

Figure 4.1   *The 70 m dish antenna at the Goldstone Deep Space Communications Complex, located in the Mojave Desert in California, is one of three complexes which comprise NASA's Deep Space Network (DSN). The DSN provides radio communications for all of NASA's interplanetary spacecraft and is also utilized for radio astronomy and radar observations of the solar system and the universe (courtesy NASA/JPL/Caltech)*

frequency, and most antennas are operated at resonance. In view of this there is only a limited band over which the antenna operates efficiently. These characteristics are governed largely by the dimensions of the antenna. The larger the antenna or more strictly the antenna elements, the lower the resonant frequency. For example, antennas for UHF terrestrial television have relatively small elements, while those for VHF broadcast sound FM have larger elements indicating a lower frequency. Antennas for short wave applications are larger still.

The bandwidth is particularly important where transmitters are concerned. If the transmitter is operated outside the bandwidth of the antenna, it is possible that damage may occur. In addition to this the signal radiated by the antenna may be less for a number of reasons.

For receiving purposes the performance of the antenna is less critical in some respects. It can be operated outside its normal bandwidth without any fear of damage to the set. Even a random length of wire will pick up signals, and it may be possible to receive several distant stations. However, for the best reception it is necessary to ensure that the performance of the antenna is optimum. Often good antenna systems will enable a receiver to pick up stations at good strength which are totally inaudible on a receiver with a poor antenna.

Polarization

It has been mentioned (in Chapter 2) that electromagnetic waves have a certain polarization. This is important because antennas are sensitive to polarization. Most antennas are linearly polarized and it is found that antennas in which the element or elements are vertical are sensitive to vertically polarized signals, and antennas with horizontal elements are sensitive to horizontally polarized signals. For optimum reception the antenna should be polarized in the same plane as the signal to be received. If it is polarized in a different plane then the received signal level is reduced. If it is at right angles (i.e. 90 degrees) to the received signal then in theory no signal is received. Similarly when transmitting the antenna will radiate a signal with a polarization in the same plane as the elements. Obviously antennas can be developed for elliptical or circular polarization.

It is found that some multi-element antennas such as the Yagi which is described later have complicated radiation patterns. They may have side lobes that have signals with a different polarization to that of the main beam. These lobes may be linearly polarized, or even elliptically polarized, and as a result signals with a different polarization may be received at a greater strength when the antenna is aligned away from the source.

Impedance

Any antenna will present a certain impedance at the point at which it is fed. This is often called the feed impedance. The impedance of an antenna results from many factors including the size and shape of the antenna, the operating frequency and its environment. The impedance is normally complex, consisting of inductive and capacitive elements as well as a resistive one. The resistance emanates from two sources. One is the resistance of the conducting element of the antenna. This is known as the 'loss' resistance and is normally kept as low as possible because any power dissipated in this resistance is lost as heat, thereby reducing the efficiency and effectiveness of the antenna. The other resistive element is known as the 'radiation resistance'. This can be thought of as a virtual resistance and it arises from the fact that power is 'dissipated' when it is radiated. There are also reactive elements to the feed impedance. These arise from the fact that the antenna elements possess inductance and capacitance. At resonance where most antennas are used, the inductive and capacitive reactances cancel one another out leaving only the resistance. However, either side of resonance the feed impedance rapidly becomes either inductive if it is operated below the resonant frequency or capacitive if it is operated above resonance.

The value of the antenna impedance is very important if the system is to operate efficiently. In order to obtain the maximum power transfer, the antenna, the transmitter or receiver and the feeder should all have the same value of resistance. This value of resistance can be calculated using Ohm's law and by knowing the values of current and voltage at the feed point of the antenna.

Gain and directivity

One important characteristic of an antenna is the way in which it is more sensitive to signals in one direction than another. This is called the directivity of the antenna. To explain some of the features about directivity it is easier to visualize the operation of the antenna when it is transmitting. The antenna will then be found to perform in the same manner when it is receiving.

Power delivered to the antenna is radiated in a variety of directions. The antenna design may be altered so that it radiates more in one direction than another. As the same amount of power is radiated, it means that more power is radiated in one direction than before. In other words, it appears to have a certain amount of gain over the original design.

The radiation of an antenna varies in all three planes and this is often termed the radiation pattern. To fully describe the performance of an antenna a full three-dimensional representation would be required, but normally this is not needed. Instead, a diagram known as a polar diagram is used to plot the performance of the antenna in a particular plane. Essentially this plots a curve around the antenna, showing the intensity of the radiation at each point. Normally a logarithmic scale is used so that the differences can be accommodated on the plot. A simple dipole

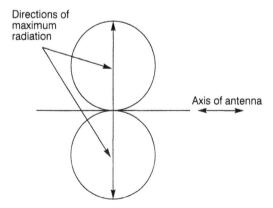

Figure 4.2 *Polar diagram of a half wave dipole*

antenna may have a pattern like that shown in Figure 4.2. From this it can be seen that the maximum radiation for transmission, and hence the maximum sensitivity to received signals, occurs when the signal is at right angles to the antenna.

When an antenna is designed to be directive, its polar diagram may look more like that shown in Figure 4.3. In this diagram, it can be seen that the antenna radiates far more signal in one direction than another. The fact that it 'beams' the power in a particular direction means that antennas of this nature are often called beams.

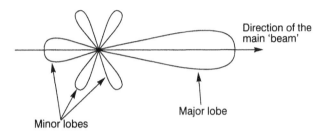

Figure 4.3   *Polar diagram of a directive antenna or beam*

The region of maximum radiation is called the major lobe. However, there are other areas around the antenna where there are significant levels of radiation. These are called minor lobes. They are always present to some extent, and generally the largest is in the opposite direction to the main lobe.

One of the major design parameters of any beam antenna is its gain. This has to be compared to another antenna. The most common antenna used for comparisons is called a dipole. The gain is simply the ratio of the signal from the beam antenna compared to the dipole expressed in decibels.

Sometimes another type of antenna may be used. Called an isotropic source it is an imaginary antenna that radiates equally in all directions. It can be calculated that a dipole has a gain of 2.1 dB over an isotropic source. When any antennas are quoted against an isotropic source the gain is measured against a dipole and then 2.1 dB is added. In view of the different gain figures obtained against the two standard antennas it is necessary to state what the gain of the antenna is being compared against in any specification. To achieve this the gain of an antenna over a dipole is quoted as a certain number of dBd (dB gain over a dipole). Similarly figures against an isotropic source are quoted as dBi (dB gain over an isotropic source).

The other factor that is important for antennas exhibiting gain is the width of the beam or beamwidth. This is the angle between the two

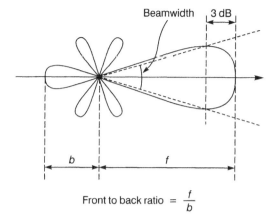

Front to back ratio $= \dfrac{f}{b}$

Figure 4.4 *The polar diagram of a typical directional beam antenna showing the beamwidth and the front to back ratio*

points on the polar diagram where the intensity falls by 3 dB from the maximum. In view of the fact that the measurement points are the −3 dB points, i.e. the half power points, it is sometimes called the half power beamwidth. As would be expected the higher the gain the narrower the beamwidth.

Gain is very important in antennas for two reasons. One is the fact that the antenna gain enables weaker signals to be received, or the transmitted power to be used more effectively giving a stronger signal where it is required. The second is the fact that the limited beamwidth means that when receiving, sources of interference can be reduced if they are coming from a different direction to the wanted signal. Similarly when transmitting it can be used to reduce the level of signal travelling in unwanted directions thereby reducing interference caused to other stations.

Another factor that is important in some applications is the front to back ratio of an antenna. This is the ratio of the maximum signal in the forward direction to the signal in the opposite direction. This figure is normally expressed in decibels. In some instances the front to back ratio is important, especially when interference in the opposite direction to the main direction is an important feature. When designing or adjusting a beam antenna it is found that the maximum gain and optimum front to back ratio do not normally coincide exactly and a compromise has to be made.

There are many different types of directive antenna and there are a number of different techniques that can be used to make an antenna directive. A simple half wavelength straight wire antenna will exhibit the maximum radiation (and sensitivity) at right angles to the axis of the antenna. By increasing the length, lobes form and the angles of main radiation tend to align more with the axis of the antenna.

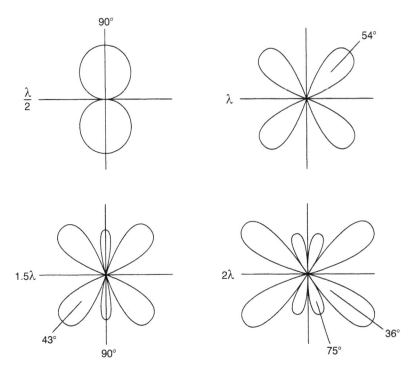

Figure 4.5   *Polar diagrams for wire antennas of differing lengths*

Directive antennas can be made by using two or more antennas. By altering the phase relationship between the signals applied to the two antennas the signal can be reinforced in one direction or another. Similarly signals can be cancelled out in some directions. It is not necessary to have to feed all the elements in a directive array. Elements placed close to the driven element will have an effect. The non-driven elements known as parasitic elements pick up the signal and re-radiate it with a different phase. Although it is not possible to have complete control over the phase and magnitude of the re-radiated signal, it is nevertheless still possible to develop highly directive antennas using this approach and the construction of the antenna is simplified by the fact that complicated phasing systems are not required. The popular Yagi antenna to be described later uses this approach.

Bandwidth

An antenna has a certain frequency bandwidth over which it can operate satisfactorily. There are two main factors that limit the performance away from centre frequency. One is the variation in impedance and the other is the changing directional pattern and beamwidth.

It has already been mentioned that the impedance of the antenna changes as the frequency changes. In turn this causes the match between a signal source and the antenna to change. When there is a poor match between the feeder used to supply the power to the antenna and the antenna, power is reflected back along the feeder. Alternatively when used for receiving, the transfer of power to the feeder will be inefficient. In both instances the efficiency is reduced although when transmitting high levels of reflected power can give rise to the possibility of damage. As a result a bandwidth dependent upon the impedance is often used for transmitting. This is usually expressed in terms of the standing wave ratio (SWR) that is explained later. For example, an antenna may operate between 55 and 60 MHz for an SWR of less than 2:1.

Changes in frequency mean that the operating point moves away from or towards the resonant frequency. With the changing values of inductive or capacitive reactance phases or current in the antenna can change and this can result in changes to the radiation pattern. Aspects like the beam width and particularly the front to back ratio are affected. For many antennas the forward gain is the main requirement and the radiation pattern bandwidth is defined as the frequency range over which the gain of the main lobe is within 1 dB of the maximum.

Angle of radiation

In many applications, another aspect associated with directivity called the angle of radiation is important. Essentially this is the radiation pattern in the vertical plane with respect to the ground. The actual angle is measured by taking the angle between the ground and the centre of the main lobe of the signal being radiated from the antenna. It is important because for many applications especially at very high frequencies and above where near line of sight communications are used. Here the power

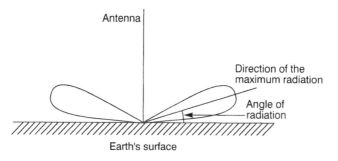

Figure 4.6 *Angle of radiation*

is required to travel parallel to the earth. In this way, it will follow the earth's surface and reach the maximum number of stations. Power directed upwards is often wasted, especially at VHF and above where this power is not usually reflected back to earth.

The angle of radiation is also very important for ionospheric propagation. A low angle of radiation will enable greater distances to be reached as a result of the geometry. However, many broadcast stations adjust the angle of radiation from their antennas to enable the signal to reach the required area for the transmission.

Antenna system

It is not possible to consider the antenna itself in isolation. It is necessary to look at the complete antenna system and optimize the operation of the whole system. The antenna system can be split down into a number of different sections. There are the antenna elements themselves. There is also a feeder that is required to transport the signal between the antenna and the transmitter or receiver. This is required because the antenna is rarely in the same place as the transmitter or receiver and a means is required to transport this power with the minimum of loss. Sometimes some form of matching arrangement is needed to ensure that the impedance match between the antenna and the feeder is optimized so that the maximum power transfer can be effected. However, in many instances the antenna is designed to match the impedance of the feeder it is to be used with.

Feeder

The purpose of a feeder is to carry radio frequency signals from one point to another with the minimum amount of signal loss. In view of the fact that radio frequency signals are being carried, ordinary wire like that used for carrying mains power is unsuitable. Feeders with suitable radio

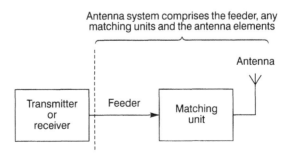

Figure 4.7 *An antenna system*

frequency characteristics are required. A poor feeder may result in the whole antenna system being degraded. It is therefore very important that a suitable feeder is used.

The operation of a feeder may not be as straightforward as might be expected at a first look. There are a number of characteristics that need to be taken into consideration.

Feeder impedance

One of the first aspects of a feeder is its impedance. Just as an antenna has a certain value of impedance, and a receiver or transmitter has an input or output impedance, a feeder has what is called its characteristic impedance. Again this is measured in ohms. The impedance is particularly important because it has to be matched to the value of impedance for the transmitter or receiver and the antenna if the optimum performance is to be achieved.

The impedance of the feeder is governed by a number of factors. The dimensions of the feeder have a large bearing on the impedance, as does the dielectric constant of the medium in and around the feeder. By controlling these factors it is possible to manufacture feeders of the required impedance.

Standing waves

When the whole antenna system is perfectly matched the maximum power transfer is obtained. When this does not happen, standing waves are set up in the feeder.

When power is transferred from a source into the load, the maximum power transfer occurs when the load and source have the same impedance and are said to be matched. In the case of a feeder and an antenna, the feeder acts as the source and the antenna is the load. If there is a poor match only a proportion of the power will be transferred from the feeder into the antenna. The remaining power from the feeder cannot just disappear and is reflected back along the feeder. When this happens the voltages and currents in the feeder add and subtract at different points along the feeder. The result of this is that standing waves are set up.

The way in which the effect occurs can be demonstrated with a length of rope. If one end is left free and the other is moved up and down as shown in Figure 4.8, the wave motion can be seen to move down along the rope. However, if one end is fixed a standing wave motion is set up, and points of minimum and maximum vibration can be seen.

When the feeder and load are perfectly matched the current and voltage will be constant along the feeder as shown in Figure 4.9.

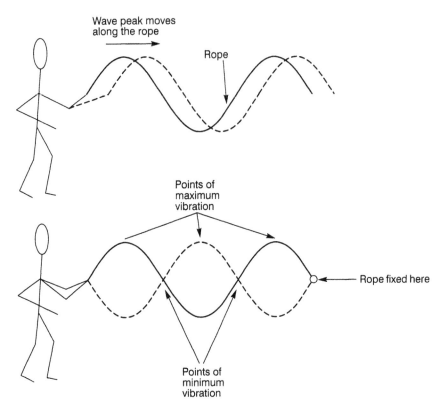

Figure 4.8 *Analogy of standing waves*

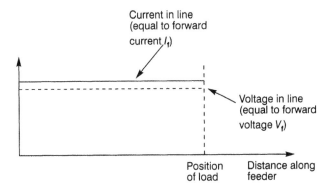

Figure 4.9 *Voltage and current magnitude along a perfectly matched line*

If the load impedance does not match that of the feeder a discontinuity is created. The feeder wants to supply a certain voltage and current ratio, while the load must also obey Ohm's law as well and cannot accept the same voltage and current ratio. To take an example, a 50 ohm feeder with 100 watts entering will have a voltage of 70.7 volts and a current of 1.414 amps. A 25 ohm load would require a voltage of 50 volts and a current of 2 amps to dissipate the same current. To resolve this discontinuity, power is reflected and standing waves are generated.

When the load resistance is lower than the feeder impedance, voltage and current magnitudes like that shown in Figure 4.10 are set up. Here the total current at the load point is higher than that of the perfectly matched line, whereas the voltage is less.

The values of current and voltage along the feeder vary as shown along the feeder. For small values of reflected power the waveform is almost sinusoidal, but for larger values it becomes more like a full wave rectified sine wave. This waveform consists of voltage and current from the forward power plus voltage and current from the reflected power. At a

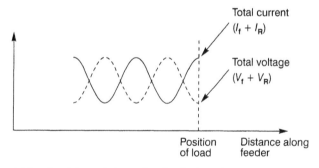

(a) Load resistance is lower than feeder impedance

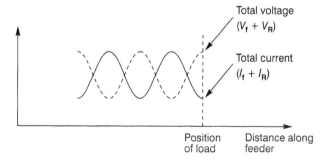

(b) Load resistance is higher than feeder impedance

Figure 4.10 *Voltage and current magnitudes for a mismatched line*

distance a quarter of a wavelength from the load the combined voltages reach a maximum value while the current is at a minimum. At a distance half a wavelength from the load the voltage and current are the same as at the load.

A similar situation occurs when the load resistance is greater than the feeder impedance; however, this time the total voltage at the load is higher than the value of the perfectly matched line. The voltage reaches a minimum at a distance a quarter of a wavelength from the load and the current is at a maximum. However, at a distance of a half wavelength from the load the voltage and current are the same as at the load.

It is often necessary to have a measure of the amount of power that is being reflected. This is particularly important where transmitters are used because the high current or voltage values may damage the feeder if they reach very high levels, or the transmitter itself may be damaged. The figure normally used for measuring the standing waves is called the standing wave ratio (SWR), and it is a measure of the maximum to minimum values on the line. In most instances the voltage standing wave ratio (VSWR) is used.

The standing wave ratio is a ratio of the maximum to minimum values of standing waves in a feeder. The reflection coefficient (ρ) can be defined and this is the ratio of the reflected current or voltage vector to the forward voltage or current vector. It is therefore very easy to calculate the SWR. The minimum value of standing wave is $(1 - \rho)$ and the maximum is $(1 + \rho)$. The standing wave ratio then becomes:

$$\text{SWR} = \frac{(1 + \rho)}{(1 - \rho)}$$

From this it can be seen that a perfectly matched line will give a ratio of 1:1 while a completely mismatched line gives ∞:1. Although it is perfectly possible to quote SWR values of less than unity, it is normal convention to express them as ratios greater than one.

Even though the voltage and current vary along the length of the feeder, the amount of power remains the same if losses are ignored. This means that the standing wave ratio remains the same along the whole length of the feeder.

Often the forward and reflected power may be measured. From this it is easy to calculate the reflection coefficient as given below:

$$\rho = \frac{P_{\text{ref}}}{P_{\text{fwd}}}$$

where P_{ref} is the reflected power and P_{fwd} is the forward power.

Loss

Another important factor about a feeder is the loss that it introduces into the system. The ideal scenario would be for the same amount of power to appear at the far end as entered from the generator. In reality the radio frequency power leaving the far end of the cable is always less than that which entered.

There are a number of reasons for this. The first is that the conductors in the feeder have a certain resistance, and as a result some of the power is dissipated as heat. To reduce this the conductors can be made thicker, but this increases the size of the whole cable, and increases the cost.

Losses are also introduced by the dielectric. A dielectric material is used between conductors in the feeder to act as an insulator and spacer to keep conductors a given distance apart. Poor quality dielectric can dissipate some power. It is for this reason that many dielectrics are semi-air spaced, consisting of plastic with air holes of one variety or another. Moisture entering the feeder dielectric can increase the loss dramatically. As a result it is necessary to ensure that any feeders which are used externally need to be well sealed where required.

Finally some power can be lost by radiation. In many types of feeder the amounts of power lost by radiation are relatively low. However, the reverse effect can be serious in some instances. If the feeder runs through an area where noise levels are high it may pick up significant amounts of interference which could adversely affect reception, despite having an antenna in a location where interference local levels are low.

The loss introduced by a certain type of feeder is proportional to its length. As a result figures for this type of specification are given for a given length of feeder, often 10 metres. They are also expressed in decibels. The frequency of operation also has a major influence, the loss rising with frequency. As a result a typical specification for a feeder might be that it has a loss of 1.0 dB at 100 MHz and 3.8 dB at 1000 MHz for a 10 metre length. For the best operation of the antenna it is necessary to ensure that the loss is minimized in keeping with the cost of installing the feeder.

Velocity factor

When a signal travels in free space it travels at the speed of light. It would also travel at the same speed in a feeder if it did not contain an insulating dielectric. The speed of the signal is reduced by a factor of $1/\sqrt{\varepsilon}$ where ε is the dielectric constant. The dielectric constant is always greater than one, and as a result the speed of the signal is always less than the speed of light. For many coaxial types of feeder the velocity factor is around 0.66

(i.e. 0.66 times the speed of light) although in some cases it can be as low as 0.5. For open wire type feeders the velocity factor may be much greater at around 0.98.

Apart from reducing the velocity of the signal, the wavelength also changes and is reduced by the same factor. The frequency remains exactly the same. This results from the fact that the lower speed means that it travels a shorter distance in the same time. In many cases this does not cause a problem, but for applications where the feeder is cut to a specific number of wavelengths, this can be crucial.

Types of feeder

There are a number of different types of feeder that can be used dependent upon the requirements of the application. Some types are most suitable for more specialist or exacting requirements whereas other types like the very familiar coaxial feeder provide good performance in a wide variety of areas.

Coaxial feeder

This type of feeder is the most commonly used, being found in most homes for television or VHF FM down leads. Apart from this it is widely used for very many other applications where a radio frequency feeder is required.

Coaxial feeder consists of two concentric conductors, spaced apart from one another by an insulating dielectric as shown in Figure 4.11. The inner conductor may be single or multi-stranded, but the outer one normally consists of a braid as shown in the diagram. The outside of the cable is covered by a protective sheath to prevent moisture ingress, as well as providing some mechanical protection.

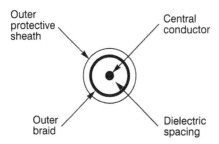

Figure 4.11   *Cross-section through a piece of coaxial cable*

Coax has a number of advantages over some other types of feeder that may be used. These include the fact that it is not affected by nearby objects. This means that it can be run almost anywhere without any undue detrimental effects. Although it will pick up or radiate very small amounts of signal this is normally small enough to be ignored. Where superior screening is required, the screen can be made up from a double layer of braid. In some instances it may be made from solid copper, although this makes it very difficult to bend.

The cable can be considered to carry currents in both the inner and outer conductors. As they are equal and opposite they cancel one another out and all the fields are confined to the cable. In fact the signal propagates along the inside of the coax as an electromagnetic wave. It is for this reason that its operation is not affected by the proximity of nearby objects.

There are a number of different types of coax. The main distinguishing feature is the impedance. For most television and domestic hi-fi antennas 75 ohms has been adopted as standard. For commercial and industrial radio frequency applications as well as amateur radio and citizens band equipment 50 ohms is used as the standard. Many computer applications use other impedances although 50 ohms is sometimes used.

The impedance is determined by the dimensions of the conductors and the dielectric constant of the material between them. It can be calculated from the formula:

$$Z_o = \frac{138}{\sqrt{\varepsilon}} \log_{10}\left(\frac{D}{d}\right)$$

where D is the inside diameter of the outer conductor or braid
d is the outer diameter of the inner conductor
ε is the dielectric constant of the material between the two conductors

Apart from the loss of a feeder that has already been covered, the other aspect of a coaxial feeder to note is that it is what is called an unbalanced feeder. This means that one of the conductors is connected to earth. As one would expect the outer braid is always connected to earth as it acts as a screen. The fact that coax is unbalanced means that it must be used with a system which can tolerate unbalanced feeders. This is normally no problem with transmitters and receivers, but some antennas require a balanced feed. In cases like this a balun must be used. This can be a radio frequency transformer used to isolate the signals from a direct connection to earth. Another method adopted on some antennas is to coil the coax, or use ferrite beads to prevent radio frequency energy propagating along the outside of the feeder.

Balanced feeder

In some instances balanced feeders need to be used in antenna systems. Often called twin or open wire feeder this type of feeder is not nearly as widely used as coax but has a number of advantages in many applications.

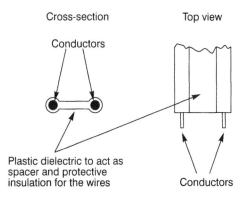

Figure 4.12 *Twin feeder*

A balanced or twin feeder consists of two parallel conductors as shown. The currents flowing in both wires run in opposite directions but are equal in magnitude. As a result the fields from them cancel out and no power is radiated or picked up. To ensure efficient operation the spacing of the conductors is normally kept to within about 0.01 wavelengths.

This type of feeder can take a variety of forms. An 'open wire' feeder can be made by having two wires running parallel to one another. Spacers are used every 15 to 30 centimetres to maintain the wire spacing. Usually these are made from plastic or other insulating material. The feeder may also be bought as a 300 ohm ribbon consisting of two wires spaced with a clear plastic. This is often used for temporary internal VHF FM antennas. It can also be bought with a black plastic dielectric with oval holes spaced at intervals. This type gives a better performance than the clear plastic varieties that absorb water if used outside.

Like coaxial cable, the impedance of twin feeder is governed by the dimensions of the conductors, their spacing and the dielectric constant of the material between them. The impedance can be calculated from the formula given below:

$$Z_o = \frac{276}{\sqrt{\varepsilon}} \log_{10}\left(\frac{D}{d}\right)$$

where D is the distance between the two conductors
    d is the outer diameter of the conductors
    ε is the dielectric constant of the material between the two conductors

Open wire or twin feeder is not nearly as widely used as coax, although it provides an ideal solution for a number of applications, especially those in the short wave part of the spectrum. It has a velocity factor of about 0.98 when the open wire version is used, and can offer very low levels of attenuation if it is kept away from other objects. The main drawback is that it is affected by nearby objects and as a result it cannot be taken through buildings in the same way as coax. This limits its use considerably, and it is rarely seen in domestic applications except for use with temporary VHF FM antennas.

Waveguide

A third major type of feeder is called waveguide. This is only used for microwave frequencies and it consists of a hollow pipe along which the signals propagate. Waveguides can be circular, but it is more common to see rectangular types as shown in Figure 4.13. They are different to the more conventional forms of feeder in that there are not two conductors. Instead the signal is introduced into the waveguide and this carries it as far as required, the walls preventing the signal escaping.

A signal can be entered into the waveguide in a number of ways. The most straightforward is known as a launcher. This is basically a small probe that penetrates a small distance into the centre of the waveguide itself as shown. Often this probe may be the centre conductor of the coaxial cable connected to the waveguide. The probe is orientated so that it is parallel to the lines of the electric field that is to be set up in the waveguide. An alternative method is to have a loop that is connected to the wall of the waveguide. This encompasses the magnetic field lines and sets up the electromagnetic wave in this way. However, for most

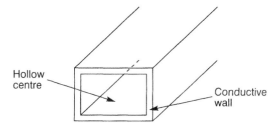

Figure 4.13   *A rectangular waveguide*

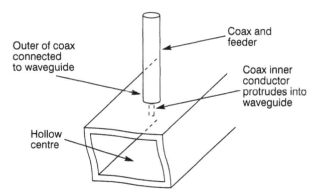

Outer of coax
connected
to waveguide

Coax and
feeder

Coax inner
conductor
protrudes into
waveguide

Hollow
centre

Figure 4.14   *A waveguide launcher*

applications it is more convenient to use the open circuit probe. These launchers can be used for transmitting signals into the waveguide as well as receiving them from the waveguide.

The dimensions of a waveguide are very important. It is found that below a given frequency called the critical frequency a waveguide will not operate. Also if the waveguide is made too large its cost will be higher than necessary and there is the possibility of higher order modes travelling along it which may introduce problems. As a result there are a variety of standard sizes of waveguide, and the correct one can be chosen dependent upon the frequency in use. They are allocated numbers and as an example waveguide WG10 is used for frequencies between 2.60 and 3.95 GHz. The advantage of a waveguide is that it offers a very low degree of attenuation or loss at these frequencies. At these frequencies the loss for a 30 metre length would only be about 1 dB dependent upon the exact frequency and whether the waveguide walls were made from aluminium or copper. This particular waveguide has internal dimensions of 72×34 mm and has a cutoff frequency of 2.08 GHz.

Signals can be transmitted directly out of a waveguide into free space. Although the directional properties are not very good there is no need to terminate them in an antenna for them to be able to radiate a signal. As a result it is very important NEVER to look down a waveguide as it is possible that power could be radiated from it if it is connected to a source of radio frequency power. High levels of power can quickly damage the eye, as people have found to their cost in the past.

Types of antenna

There is a wide variety of different types of antenna that can be used and each has its own advantages and may be used for specific applications. Often an antenna may be envisaged in a format similar to a television array, alternatively it may be a wire antenna, or it may just be a short length of metal like those seen on automobiles. Some may be quite sophisticated and give higher levels of performance, but at a higher level of cost and possibly larger in size. Simple antennas may also be able to perform a satisfactory function. For each application it is necessary to understand the requirements and use this information to select the optimum type. Gain, directivity, frequency of operation, bandwidth, efficiency and many more attributes all need to be considered when making the choice of antenna. In addition to this some may be horizontal while others may be vertical, either for operational considerations, or possibly for mechanical considerations. As a result a vast number of different types of antenna may be seen for differing applications.

The simplest type of antenna is one that is very short compared to a wavelength. However, these antennas are very inefficient and seldom used. Possibly the most common type of antenna is the dipole. This can be used on its own or as a building block in a more complicated antenna.

Dipole

The most widely used form of the dipole is the half wave version, although it can be any odd multiple of electrical half wavelengths. The half wave version consists of a half wavelength section of wire, with the feed point in the centre of the antenna as shown in Figure 4.15.

To understand more about the operation of the antenna it is necessary to look at the voltage and current distribution along the wire. From Figure 4.16 it can be seen that they vary sinusoidally along the length of the antenna. The voltage has a maximum at either end, and a minimum

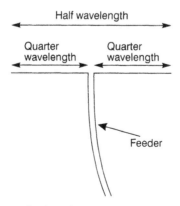

Figure 4.15 *A half wave dipole antenna*

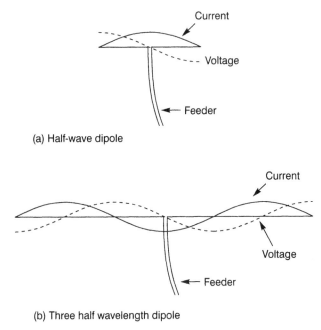

(a) Half-wave dipole

(b) Three half wavelength dipole

Figure 4.16   *Voltage and current distribution of dipole antennas*

in the middle where the feed point is taken. The current follows the opposite pattern, falling to a minimum at the end and having a maximum in the middle. Longer dipoles follow the same basic principles having further wavelengths. However, a voltage maximum and current minimum is found at both ends. The feed point is similarly taken at a point of voltage minimum and current maximum. A three half wavelength antenna is sometimes used, and this could be fed in the middle, or at one of the other current maximum points.

With the feed point taken at the point where the voltage is a minimum and the current a maximum it can be understood from Ohm's law that the impedance is low. The actual impedance depends mainly on the proximity of nearby objects, but in free space the value of the impedance is 73 ohms, making it a good match to 75 ohm coax.

It is very easy to alter the impedance of a dipole. The proximity of nearby objects has a major effect. The height above ground alters the impedance quite significantly. It is possible to calculate this effect and plot the impedance of a dipole in terms of its height in wavelengths above ground as shown in Figure 4.17.

In some antenna designs where the dipole is used as the basic driven element, the feed impedance of the dipole falls to very low values, often to values of 10 ohms or less. This can be very difficult to feed efficiently.

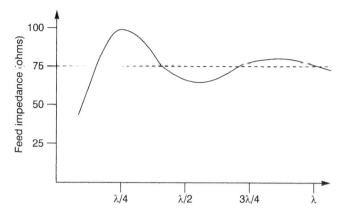

Figure 4.17 *Approximate impedance of a horizontal dipole at varying heights above ground*

The impedance can be increased to more suitable values by using what is called a folded dipole. In this type of antenna an additional wire is taken from one end to the other. Different conductor sizes can be used for the original section of the dipole element and the folded section. If they are both the same then the impedance of the dipole is raised by a factor of four. By changing the ratios of the sizes of the conductors it is possible to tailor the impedance. However, in most cases the same sizes are used, giving a basic impedance of 300 ohms for a folded dipole (i.e. 4 × 75 ohms).

An additional advantage of using a folded dipole is that it gives an increased bandwidth over a basic dipole. This is of great advantage when the antenna needs to be used over a band of frequencies as in the case of VHF FM or UHF television broadcasts.

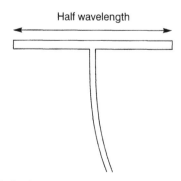

Figure 4.18 *A folded dipole*

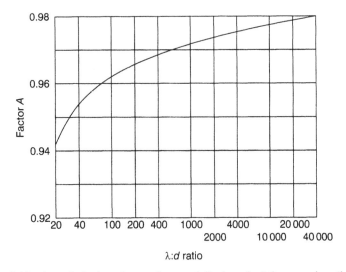

Figure 4.19   *Length factor of an antenna plotted against the wavelength, element diameter ratio*

The length of the dipole is of prime importance because it determines the resonant frequency. This can be determined quite easily although it is not quite the same as a half wavelength in free space. Instead it is slightly shorter. This is due to a number of effects including one referred to as the end effect. This is largely dependent upon the length to diameter ratio of the wire or conductor that is used. In most cases the reduction in length is around 5 per cent, although it is possible to determine a more exact value from the graph in Figure 4.19.

From a knowledge of the length reduction factor it is possible to calculate the antenna length in either metres or inches:

$$\text{length (metres)} = \frac{A \times 150}{f}$$

$$\text{length (inches)} = \frac{A \times 5905}{f}$$

where f is the frequency of operation in MHz.

When making a first version of an antenna it is always wise to cut the elements slightly longer than the calculated length. This enables the length to be reduced when making alterations to reach the optimum performance. It is always more difficult to add wire or metal that has been previously removed.

End fed wire

One of the easiest types of antenna to install for short wave applications is an end fed wire. Often misnamed a long wire, this type of antenna can be erected very easily and can be used on a variety of frequencies. As such these antennas are used in a number of applications including ships, aircraft as well as being very convenient for short wave listeners and radio amateurs.

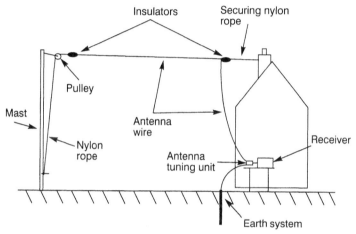

Figure 4.20 *An end fed wire antenna system*

The antenna is normally made to be a quarter wavelength or longer at the lowest frequency of operation, but to give a good match it is necessary to use the antenna with a matching unit. It also needs to be operated with a good earth system. A poor earth will reduce the efficiency of the system.

The directional pattern of the system will depend upon the length of the antenna. A quarter wavelength system will have its maximum sensitivity at right angles to the axis of the wire. Longer ones will have lobes which move progressively towards the axis of the wire itself. If the antenna is made several wavelengths long it becomes what is known as an 'end fire' antenna where the main lobes are almost in line with the axis of the wire. This is what a true long wire is.

Phased arrays

It has already been mentioned that phasing techniques can be used to create a directive antenna. By changing the phase between two driven

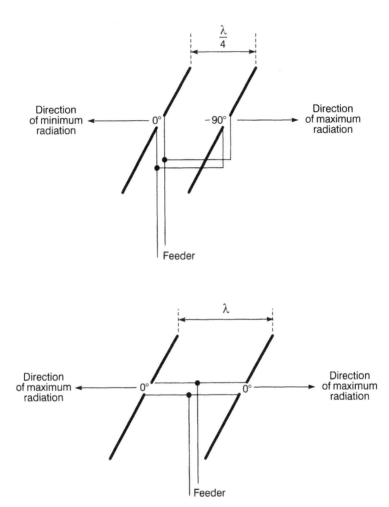

Figure 4.21   *Phased antennas using two dipoles*

elements the signal can be cancelled out or reinforced in various directions. There is a wide variety of possibilities that can be created, but to demonstrate the principle take the simple example of two dipoles spaced a quarter wavelength apart. If they are fed 90 degrees out of phase then the result is that the signals from the two antennas are in phase in one direction and they cancel because they are out of phase in the other. A bi-directional pattern can be obtained by feeding the antennas in phase with one another and spacing them one wavelength apart. In this way the signals from both antennas reinforce one another as shown in the diagram.

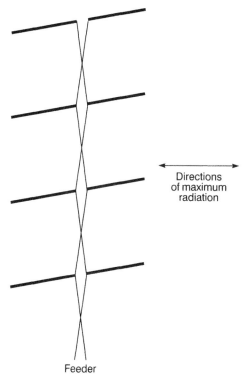

Feeder

Figure 4.22 *A broadside array*

Another type of phased antenna is known as a broadside array. Here the antenna is arranged in such a way that the maximum radiation occurs in a plane perpendicular to the plane containing the driven elements. To achieve this all the elements must be fed in phase with one another. This is achieved using a phasing harness. In the simplest case the dipoles are spaced as shown. An electrical half wavelength of feeder is used between each of the dipoles and to keep the phase at the fed point the same a phase reversal is introduced. The directional properties of a broadside array depend on the number of elements used, and whether the elements themselves have gain.

Yagi

The Yagi is certainly the most popular form of directional antenna. More correctly named the Yagi–Uda after its Japanese inventors it was first outlined in a paper presented in 1928. Nowadays the antenna is widely used where a directional array is required, being almost the only type of antenna used for UHF television reception.

The basic antenna consists of a central boom with the elements mounted to it at right angles as shown. The antenna consists of the main driven element to which the feeder is connected, and parasitic elements either side. These parasitic elements are not directly connected to the feeder but operate by picking up and re-radiating power in such a phase that the directional properties of the antenna are altered. This is achieved having the phase of the current in the parasitic element or elements in such a phase that it reinforces the signal in a particular direction, or cancels it out.

There are two main types of parasitic element: reflectors that reflect power back towards the driven element, and directors that increase the power levels in the direction of the directors. The properties of a parasitic element are determined by their spacing and their electrical length.

When a parasitic element is made inductive the induced currents are in such a phase that they re-radiate power back towards the driven element, i.e. reflecting the power. To make an element into a reflector it is tuned slightly below resonance. This can be done by physically adding a coil to provide the additional inductance, or more usually by making it physically slightly longer. Typically reflectors are made about 5 per cent

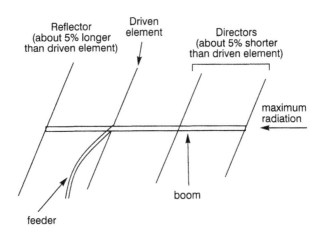

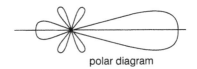

polar diagram

Figure 4.23 *A Yagi antenna*

longer than the driven elements. Conversely directors are made by tuning the element above resonance. This can be done by adding some capacitance into the element, or making it slightly shorter than the driven element, again by about 5 per cent.

Only one reflector is used, as the addition of further reflectors behind the main one adds very little to the performance. Further directors do give increased directivity, and it is not uncommon to see ten or more directors on an antenna. Normally the number is limited by size, cost and the required amount of gain. Typically a two element design consisting of a driven element and reflector or director will give around 5 dB gain over a dipole and a three element antenna with a director as well gives around 7 dB. Additional directors give less gain, starting at around 2 dB for the first one or two; however, as a rough guide each additional director gives around 1 dB. For example, a typical 11 element array is likely to have a gain of around 13 dB and a 12 element version, around 14 dB. It is also found that the element spacing has a small effect on the gain. Usually the spacing is between 0.1 and 0.3 wavelengths.

The polar diagram of a Yagi is generally like that shown in Figure 4.23. As more elements are added the beamwidth becomes much narrower. This makes positioning the antenna more critical, a point to be remembered when choosing the required gain for an antenna.

The feed impedance is altered by the parasitic elements. In the same way that the presence of the earth lowers the feed impedance of the

Figure 4.24 *A Yagi used for television reception*

dipole, placing parasitic elements close to the driven element lowers its impedance. The element spacing has a larger effect on the feed impedance than it does on the gain. In most designs a folded dipole is used and the spacing adjusted to give the optimum match.

As there is only one driven element and the remainder are parasitic elements it is not possible to achieve complete control of the phasing and magnitude of the currents in all the elements. As such it is not possible to have complete control over the radiation pattern. In view of the large number of variables involved in the design, and the compromises that need to be made, computer programs are widely used in the design of these antennas. This enables their performance to be accurately assessed before any antennas are made. As a result there have been considerable performance improvements in Yagi antennas in recent years.

Discone

The discone is often used where a wideband receiving antenna is needed. For example, it is very popular with scanner enthusiasts, and it is also used in many commercial applications. It is almost omnidirectional and it can operate over a frequency range of up to 10:1 in certain instances. In addition to this it offers a low angle of radiation over most of its operating range. It also presents a relatively good match to the feeder over its operating range, although there is obviously some variation, particularly at the extremes of the range.

The discone derives its name from its distinctive shape. Looking at Figure 4.25 it can be seen that the antenna basically consists of a disc section and a cone section that are simulated by a number of rods. The disc section is insulated from the cone by a block of material that also acts as a spacer keeping the two sections a fixed distance apart. In fact this distance is one of the factors that determines the overall frequency range of the antenna.

When designing a discone the length of the cone elements should be a quarter wavelength at the minimum operating frequency. The disc elements are generally made to have an overall length of 0.7 of a quarter wavelength. The diameter of the top of the cone is mainly dependent upon the diameter of the coaxial cable being used. The spacing between the cone and the disc should be about a quarter of the inner diameter of the cone. Making the minimum diameter of the cone small will increase the upper frequency limit of the antenna.

Operation of the antenna is fairly complicated but it can be visualized in a simplified manner. The elements that form the two sections of the antenna electrically simulate a complete disc and cone from which the energy is radiated. Naturally the more elements that are used the better the simulation of the cone, but typically a good balance between cost and

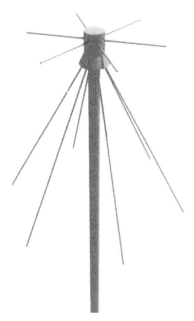

Figure 4.25 *A discone antenna*

electrical performance is achieved when six or eight elements are used. Addition of further elements would also add to the wind resistance of the antenna.

In operation the energy from the feeder spreads out over the surface of the cone towards the base of the antenna until a point on the cone is reached where the vertical distance between the point on the cone and the disc is a quarter wavelength. At this point 'resonance' is seen and the energy is radiated.

The radiated signal is vertically polarized and the radiation pattern is very similar to that of a vertical dipole. Although some variation is seen over the operating band particularly at the top, it maintains a very good low angle of radiation over most of the range.

It is found that the current maximum is at the top of the antenna and that below the minimum frequency the antenna presents a very bad mismatch to the feeder. However, once the frequency rises to within the operating band a good match to 50 ohm coax is maintained over virtually the whole of the operating range.

Log periodic

The log periodic antenna is a wideband antenna that is directional and capable of providing significant levels of gain and operating over a frequency range of up to about 2:1. Developed in 1955 at the University of Illinois it has found widespread acceptance in many areas including the military.

The log periodic antenna has many similarities to the more familiar Yagi because it exhibits forward gain and has a significant front to back ratio. In addition to this the radiation pattern stays broadly the same over the whole of the operating band as do parameters like the radiation resistance and the standing wave ratio. However, it offers less gain for its size than does the more conventional Yagi.

The most common form of log periodic antenna is known as the log periodic dipole array (LPDA). The basic format for the array is shown in Figure 4.26. Essentially it consists of a number of dipole elements of a size that steadily diminishes from the back of the beam where the largest element is a half wavelength at the lowest frequency. The element spacing also decreases towards the front of the array where the smallest elements are located. In operation there is a smooth transition along the array of the elements that form the active region as the frequency changes. To ensure that the phasing of the different elements is correct, the feed phase is reversed as shown in the diagram.

The operation of a log periodic antenna can be gained by taking the example of operation around the middle of the operating band. It can be seen that there is a phase reversal between elements in the antenna and this is crucial to its operation. When the signal appears on the antenna itself it first sees the smaller elements at the 'front'. These are spaced quite close together in terms of the operating wavelength and this means that the fields from these elements cancel one another out as the feeder sense is reversed between the elements. Then as the signal progresses down the

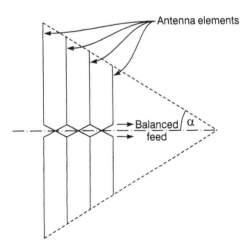

Figure 4.26   *A log periodic array*

antenna a point is reached where the feeder reversal and the distance between the elements give a phase shift of about 360 degrees. At this point the effect which is seen is that of two phased dipoles. The region in which this occurs is called the active region of the antenna. Although the example of only two dipoles is given, in reality the active region can consist of more elements. The actual number depends upon the angle α and a design constant. The elements that are not 'active' still contribute to the operation of the antenna. The larger elements behind the active region are longer than the resonant length and appear inductive. Those in front are shorter than the resonant length and appear capacitive. These are exactly the same criteria that are found for reflectors and directors in a Yagi. Accordingly those elements behind the active region act to reflect the power they receive and those in front act as directors. This means that the maximum radiation is towards the feed point of the antenna.

The feed arrangements of the log period antenna are important. The input impedance is dependent upon a number of factors. Fortunately the overall impedance can be determined to a large degree by the impedance of the inter-element feeder. However, the main problem to overcome is that the impedance will vary according to the frequency in use. Fortunately this can be compensated to a large degree by making the longer elements out of a larger diameter rod. Even so the final feed impedance does not normally match a convenient 50 ohms on its own. It is normal for some further form of impedance matching to have to be used. This may be in the form of a stub or even a transformer. The actual method employed will depend to a large degree on the application of the antenna and its frequency range.

A log periodic antenna provides modest levels of gain when compared to a Yagi. Typically this is between about 4 to 6 dB and over an operating frequency range of about 2:1. As such it enables wideband operation to be achieved with some gain and the use of only a single feeder. This is a significant advantage in many applications, especially at VHF and UHF.

Vertical antenna

Vertical antennas are widely used, especially at VHF and UHF. As the antenna element is in the vertical plane, power is radiated in all directions around it. Coupled to this the antenna has a low angle of radiation, meaning that little signal is wasted by radiation in an upwards direction away from the earth. In view of their radiation pattern these antennas are ideal for mobile applications. Their all-around radiation pattern means that the antenna does not need to be repositioned as the vehicle moves. The use of these antennas is not restricted to mobile stations. Fixed

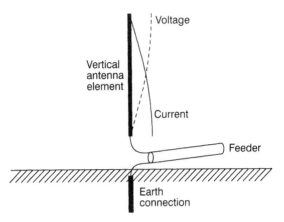

Figure 4.27   *A quarter wave vertical antenna*

stations too can use them when an omnidirectional pattern is required. Most medium wave broadcast stations use vertical antennas to enable them to cover all directions away from the antenna.

Although a vertical antenna can be made a variety of different lengths, they are often made a quarter wavelength long. This means that they are fed at a current maximum point as shown in Figure 4.27.

A vertical is an unbalanced antenna and is fed with coaxial feeder as shown. The centre conductor of the feeder is connected to the vertical element, while the screen is connected to a ground. The grounding system must be very good. A single spike into the earth may have a resistance of 100 ohms or more in some cases. As the feed impedance of a vertical of this type may be only 35 ohms, the earth resistance easily becomes the dominating factor absorbing most of the power supplied to the system. Broadcast stations using vertical antennas require the optimum performance and will ensure that a very good earth system is installed along with the antenna itself. Apart from a good earth spike at the base of the antenna radial systems consisting of wires a quarter wavelength or more are buried just below the surface of the ground.

For mobile applications the antenna is mounted onto the bodywork of the vehicle. In this case the metalwork of the vehicle acts as the ground. This is normally a very efficient grounding system, especially at VHF and above where the bodywork extends for several wavelengths away from the base of the antenna. The other advantage is that the antenna can be mounted relatively high up. The best position from a radio performance standpoint is the centre of the roof of the vehicle; however, this is not always acceptable aesthetically. In this case mounting points on the wing usually give good performance as well.

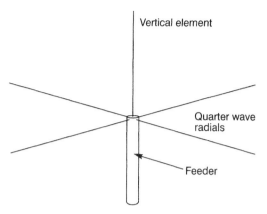

Figure 4.28 *A ground plane antenna*

It is not always convenient to mount a vertical on the ground. It can easily be masked by nearby objects reducing its effectiveness. This is particularly true for VHF and UHF applications where the antenna is physically very small and mounting it on the ground would considerably reduce its effectiveness. In situations like this a ground plane system can be used. The ground plane is made out of a number of rods or wires which are normally a quarter wavelength long. Four rods are usually employed as being sufficient for most applications and a ground plane system may look like that shown in Figure 4.28.

The feed impedance of a ground plane is low. To improve the match to 50 ohms the radials can be bent downwards. As the radials are bent further down, the antenna becomes more like a vertical dipole and its impedance increases towards a value of 73 ohms, assuming that there are no nearby objects. Alternatively the vertical element can be folded in the same way as a dipole, or a matching network can be used at the feed point to ensure a good match.

Short antennas

It is often necessary to design an antenna that is shorter than a quarter wavelength. With electronic equipment becoming smaller the same is often needed for any associated antennas. This is particularly true for cellular telephones where equipment is very much smaller than it was a few years ago, and also the antennas used on vehicles are required to be smaller.

There are a number of ways of reducing the physical size of an antenna. One method is to make the antenna electrically the right length even

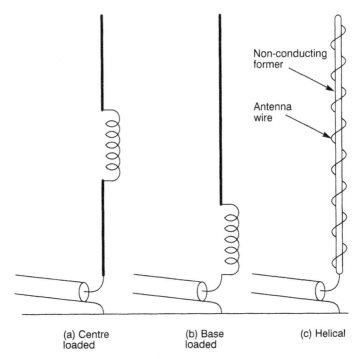

Figure 4.29   *Adding inductive loading to an antenna*

though it is physically small. It is found that when an antenna is too short for its resonant frequency it becomes capacitive. The capacitive component can be removed by adding inductance to the antenna to bring it to resonance. This can be achieved by adding a coil to a section of the antenna as shown in Figure 4.29, or by making what is called a helical antenna and distributing the inductance over the whole length. When a single coil is used to give the inductance, this is either placed in the centre as shown or at the base.

Another alternative is to make the antenna electrically short and feed it accordingly. If this method is adopted arrangements need to be made to match the impedance of the antenna to the feeder, as the two will not be matched.

Parabolic reflector

One type of antenna which has been seen far more in recent years is the parabolic reflector or 'dish'. It is widely used for frequencies above 1 GHz where very high levels of gain can be achieved. Initially these antennas

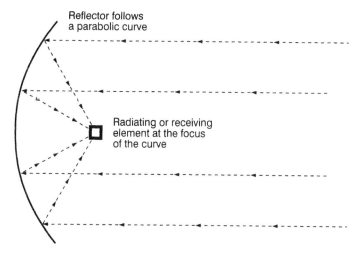

Figure 4.30 *A parabolic reflector*

were only used for professional applications, especially radio astronomy or satellite communications. However, with the advent of satellite television these antennas are often seen on the sides of houses for reception of these broadcasts.

The antenna consists of a radiating element that may be a simple dipole or a waveguide horn antenna. This is placed at the focal point of the parabolic reflecting surface. The energy from the radiating element is arranged so that it illuminates the reflecting surface. Once the energy is reflected it leaves the antenna system in a narrow beam. As a result considerable levels of gain can be achieved.

The gain is dependent on a variety of factors, but is mainly a function of the diameter of the parabolic reflector. The actual gain can be estimated from the formula:

$$G = 10\log_{10} k \left(\frac{\pi D}{\lambda}\right)^2$$

where G is the gain over an isotropic source
  k is the efficiency factor which is generally about 50 per cent
  D is the diameter of the parabolic reflector in metres
  λ is the wavelength of the signal in metres

From this it can be seen that very large gains can be achieved if sufficiently large reflectors are used. However, when the antenna has a very large gain, the beamwidth is also very small and the antenna requires very careful control over its position. In professional systems electrical servo systems are used to provide very precise positioning.

The reflecting surface may not be as critical as may be thought at first. Often a wire mesh may be used. Provided that the pitch of the mesh is small compared to a wavelength it will be seen as a continuous surface by the radio signals. If a mesh is used then the wind resistance will be reduced, and this may provide significant advantages.

Ferrite rod antennas

The ferrite rod antenna is almost universally used in portable transistor broadcast receivers as well as many hi-fi tuners where reception on the long, medium and possibly the short wave bands is required. As the name suggests the antenna consists of a rod made of ferrite, an iron-based magnetic material. Coils are wound around this as shown in Figure 4.31.

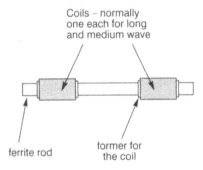

Figure 4.31 *A ferrite rod antenna*

This coil is brought to resonance using a variable tuning capacitor contained within the radio circuitry itself and in this way the antenna can be tuned to resonance. As the antenna is tuned it usually forms the RF tuning circuit for the receiver, enabling both functions to be combined within the same components, thereby reducing the number of components and hence the cost of the set.

The ferrite rod antenna operates using the high permeability of the ferrite material to 'concentrate' the magnetic component of the radio waves as shown in Figure 4.32. This means that the antenna is directive. It operates best only when the magnetic lines of force fall in line with the antenna. This occurs when it is at right angles to the direction of the transmitter. This means that the antenna has a null position where the signal level is at a minimum when the antenna is in line with the direction of the transmitter.

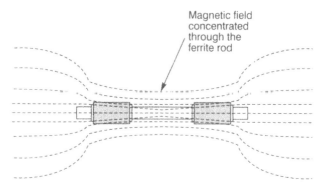

Figure 4.32 *Operation of a ferrite rod antenna*

These antennas are very convenient for portable applications, but their efficiency is much less than that of a larger antenna. The performance of the ferrite also limits their frequency response. Normally they are only effective on the long and medium wave bands, but they are sometimes used for lower frequencies in the short wave bands although their performance is significantly degraded, mainly because of the ferrite.

Loop antennas

Another form of antenna that can be used to good effect under some circumstances is a loop antenna. These antennas are formed by creating a loop of one or more turns of wire that form a short circuit to direct current. An example of this type of antenna is often seen with hi-fi systems where a small loop is used to provide the reception for long and medium wave stations.

Loop antennas fall into two categories, small loops and large loops. As the names imply the category depends on their size. As a rule, small loop antennas are those that have a circumference of less that 0.1 wavelengths. These antennas are generally only used for receiving as their radiation resistance is extremely low and losses when compared to the resistance of the wire can be high. This makes them quite inefficient, although very useful for receiving signals in areas of high signal strength.

Loop antennas are directional. A small loop is least responsive along its axis, and most responsive in the plane perpendicular to the axis of the antenna. Loop antennas generally have several turns, and often a variable capacitor is added to resonate the antenna and improve its performance. This is normally done if the antenna forms part of a complete receiver, otherwise they are often operated in a non-resonant mode where they provide a wideband response that is fairly flat.

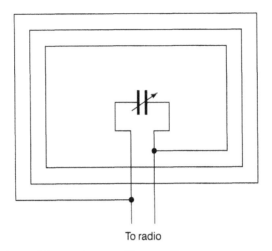

Figure 4.33   *A small loop antenna with capacitor resonance*

Large loops are classified as those with a circumference of greater than 0.1 wavelengths. Typically they have a circumference of half or a full wavelength and this makes them self-resonant without the need for external tuning. A half wave loop presents a high impedance to the feeder, but the full wave loop has a much lower impedance, of the order of 75 ohms. The maximum radiation occurs at right angles to the axis of the antenna.

Active antenna

In a number of receiving applications active antennas are used. As the name implies the antennas employ an active or amplifying element. They consist of an electrically short antenna element. An active amplifier at the base of the antenna element is used to amplify the signal and act as a matching circuit to ensure a good match to the feeder. This is normally achieved easily because a short antenna will present a high impedance and this can usually be accommodated very easily with a semiconductor amplifier.

Power for the amplifier is normally supplied along the coax from a power supply at the receiver end of the feeder. The centre conductor of the coax allowing the supply to be passed to the amplifier, while also carrying the signal to the receiver. In this way it is possible to have the power source near the receiver without the need for additional wires to be routed to the antenna.

The design of active antennas is not easy. If they are to perform well they must be capable of good strong signal handling over a wide frequency range. This is necessary because the antenna is untuned and will pick up signals over a wide frequency band, some of which will be very strong. To achieve the required performance devices capable of handling relatively high powers are used. Poor performance will result in intermodulation products being generated, and the receiver seeing many more signals than the receiver element is actually picking up.

Practical aspects of installing antennas

The practical aspects of antenna installation are particularly important. The performance of the antenna can be dependent as much if not more on the location and the way the antenna is installed as on the design of the antenna itself. It is therefore worth while spending some time and space to outline some aspects of antenna installation.

Location

There are many types of antenna installation that can be seen. They range from small domestic antennas for applications such as broadcast television of radio to large commercial installations for broadcast, satellite or one of a variety of applications. Whatever the application the location can play a large part in the performance of the antenna. For most people the installation of an antenna will be associated with a domestic installation, although the same basic concepts apply to any antenna installation.

Height is an important consideration. For optimum operation the antenna should be above any local objects that might tend to screen the antenna. This is particularly important where low angles of radiation are required. It is also found that additional height also tends to reduce the effect of local interference, reducing any electromagnetic compatibility (EMC) problems that may occur.

For many domestic installations it is necessary to have the antenna mounted internally. Although this is not ideal it is still possible to achieve satisfactory levels of performance in many instances. The structure of the house will introduce a degree of attenuation, but if it is situated in a high signal level area this may not be a problem. It is difficult to assess the exact level of attenuation because it depends on a variety of factors including the nature of the materials used in the house, the frequencies involved and their proximity to the antenna. For example, it is found that the level of attenuation increases when the roof is wet. There may be a 6 dB increase in attenuation at frequencies of the order of 500 MHz after rain, and this may be maintained for many hours after the precipitation has finished if the roof tiles are absorbent. Another point to watch when mounting an antenna internally is that it is kept away from internal wiring and metalwork that might screen the antenna or affect its operation by detuning it. Often wiring or metal objects such as water tanks can have a significant effect.

Another popular place to mount an antenna is on a chimney stack. This can be an ideal location as it is relatively high. However, check that any antenna mounted on the chimney does not overhang a neighbour's property. It is also necessary to check that the chimney will be able to withstand the antenna, particularly when it is windy. This is particularly important for large antennas.

In some instances masts or towers may be required. Many varieties of masts and towers are available. Some of these are free standing while others need to have guys attached. Some can be wall mounted, using the side of the house or other building for support. Towers are ideal in many respects but they are more costly than masts using guys. However, a mast with guys will take up more space. In addition to this care and expertise are required when installing them.

A further requirement for a tower or mast is that it should have lightning protection installed. A very good earth connection is required along with wide copper straps from the mast to the earth connection. In this way current flowing to the equipment will be reduced as far as possible and damage to the equipment and the area where it is installed will be minimized. In view of the importance of this protection professional advice should be sought.

Materials for antennas

The choice of materials used in an antenna system can have a marked effect on its mechanical and electrical performance. Not only does rain and other naturally occurring gases in the atmosphere cause corrosion to set in, but in many areas of the globe, the atmosphere contains sufficient amounts of pollution to further increase the levels of corrosion.

The use of dissimilar metals will cause a considerable amount of trouble as a result of electrolytic action. This arises because each metal has an electropotential. If two different metals are used then the difference in potential will result in corrosion, even when the antenna is dry. Although when moisture is present the problem is made much worse, especially when pollution is present.

Different combinations of metals react with one another to differing degrees. This results from their relative positions in the electrochemical series. The further apart the two metals are the greater the level of corrosion that is to be expected. Those much closer in the series will react less. Those at the top of the series are termed anodic metals whereas those at the bottom are termed cathodic. It is found that metals lower in the series will cause those higher in the series to corrode. For example, brass or copper screws will cause metals such as aluminium that might be used in a mast to corrode. The use of cadmium plated screws would result in far less corrosion. If different metals must be used together then metals as close as possible in the electrochemical series should be chosen, and all moisture excluded to reduce the level of corrosion.

Anodic
  Magnesium
  Aluminium
  Duralumin
  Zinc
  Cadmium
  Iron
  Chromium iron alloys
  Chromium nickel iron alloys
  Soft solder tin lead alloys
  Tin
  Lead
  Nickel
  Brasses
  Bronzes
  Nickel copper alloys
  Copper
  Silver solders
  Gold
  Platinum
Cathodic

The most obvious problem that arises from corrosion is the possible reduction in mechanical strength of an antenna, and the problems that arise when trying to disassemble an antenna system that is corroded for maintenance. However, there are other effects that can arise. Many antenna elements are made from aluminium. Also most feeders are made from copper and this means that the electrical joints made between the two can corrode, resulting in a high resistance connection that will significantly degrade the performance of the antenna. Another problem that can occur is often called the 'rusty bolt effect'. It occurs because sometimes the corroded materials can act as a semiconductor, producing a non-linear component that may generate harmonics and inter-modulation products if very strong signals are present. It is usually more applicable to transmitting stations that carry several transmissions. Here the spurious signals generated may cause interference to other local users.

Safety

An antenna can be a potentially hazardous item, both in its installation and during its use. Fortunately there are generally few accidents, although there have been fatalities as a result of their use and installation. However, with a little care and common sense it is most unlikely that any accidents will occur. The main requisite to prevent any accidents is a general awareness of safety.

Safety precautions take many forms. One of the first is that under no circumstances should an antenna be erected anywhere that it may fall onto power lines, or that power lines could fall onto it. This may seem like a remote possibility, but there have been instances of antennas falling onto power lines.

At all times the proper materials and fixings should be used. It is not worth taking short cuts as falling antennas can cause serious injury, not to mention the damage to the antenna and other property. For smaller antennas normal television fixings can often be used, but for larger or more specialist antennas seek professional advice. It is well worth investigating the wind loading likely to be encountered and whether the method used for mounting the antenna is adequate.

Care should also be taken when planning the installation of an antenna. Sufficient thought put in at the early stages can enable the antenna to be installed easily and safely. There are many considerations: the location of the antenna, the materials to be used, the way in which it is to be installed, advice to be sought, assistance to be sought, etc. In this way as many problems as possible can be addressed before they occur. As antennas are often installed high in the air, it is necessary to be confident that the installation can be undertaken safely, otherwise it is always best to call in professional assistance. Above all do not take any risks.

With adequate preparation, the right materials, sufficient planning, the right level of experience, and above all a keen awareness for safety, no problems and accidents should arise.

5 Receivers

Radio receivers are very common these days. They are used in a wide variety of applications and in many different forms. The most obvious type of receiver is the domestic broadcast radio. Today most households have a variety of sets, ranging from portable radios up to hi-fi tuners and car radios are also very common. Virtually all new automobiles these days come with a sophisticated radio already fitted. However, they are not the only types of sets to be found in common use. Cellular phones are very common and their use is increasing all the time, and naturally these contain a receiver. Receivers are also needed for the computer wireless local area network communications systems. There is also a wide range of commercial uses of radio receivers. They are used in communications systems ranging from short-range VHF or UHF walkie-talkies to the longer-range systems needed for aircraft of maritime communications.

Figure 5.1 *A typical domestic portable receiver*

Whatever the use of the receiver, the same basic principles apply and the same basic building blocks are used.

A receiver performs two main functions. The first is to remove the modulation from the radio frequency signal. This process is called demodulation or detection and gives signals at audio frequencies that can be amplified by an audio frequency amplifier and passed into head-phones or a loudspeaker. The other is to provide selectivity. With the vast number of stations on the radio frequency bands today it is necessary that the receiver is able to tune in the station on the required frequency and reject the others. If this function is not performed well then a number of different stations will be received at one time making it very difficult to copy any of them.

Amplification is another important function that radios provide. The signals picked up by the antenna are very weak, and need to be amplified to a large degree if they are to be heard in a loudspeaker or passed to some other form of unit like a modem where data is extracted.

There are a number of different methods of performing some of these functions. As a result there are a number of different types of format for receivers. Obviously some are much simpler than others. The simplest is a crystal set. This is the most basic type of receiver, and while it is not used for any serious listening these days it shows some of the basic principles required for a radio receiver.

Crystal set

The crystal set has been in existence for very many years, often being built by hobbyists these days as a first construction project. The simple circuit of a very basic set is shown in Figure 5.2, and from this it can be seen that very few components are used. Naturally the simplicity reflects the level of performance which can be achieved, and more complicated and effective designs can be built if required.

In the circuit L_1 and VC_1 act together to form a resonant circuit. This performs the function of accepting signals on and around the wanted one and rejecting the others. As the inductor and variable capacitor form a parallel tuned circuit, the impedance reaches a maximum at resonance. This means that signals on frequencies either side of the resonant frequency pass to earth. At resonance the signals do not pass through the tuned circuit. Instead they pass into the diode. This rectifies the signal so that the amplitude variations from the signal can be obtained. The small smoothing capacitor C_1 acts to remove any remaining radio frequency components and smoothes the signal as shown in Figure 5.3.

The rectified signals from the diode can be heard by connecting a pair of headphones to the output of the set. However, it is soon discovered

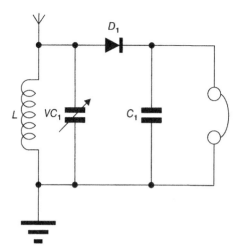

Figure 5.2   *A circuit of a crystal set*

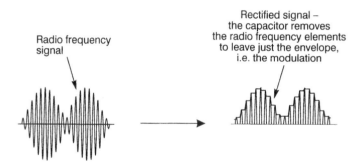

Figure 5.3   *Demodulating an AM signal*

that the signal strengths are weak even when a good antenna is used. Also the tuning is very broad and a single, strong station may be heard over a large part of the tuning range.

TRF receiver

If a crystal set is built, its limitations very quickly become obvious. The low signal strengths can be overcome by adding amplification. This can be placed after the detector as an audio amplifier, but it can also be added in the radio frequency stages prior to the detector. A receiver of this nature is called a tuned radio frequency (TRF) receiver because the tuning takes place at the radio frequency. This distinguishes it from other types mentioned later.

The lack of selectivity in the crystal set is mainly caused by the fact that there is only one tuning circuit. To overcome this problem and increase the gain still further a circuit known as a regenerative detector can be used. This type of circuit was very popular in the early days of radio when the number of valves had to be kept to a minimum to keep the cost of circuits down to what people could afford.

A regenerative detector works by feeding back some of the signal from the output to the input. A control called a regeneration or reaction control is used to keep the circuit just below the point of oscillation. At this point the maximum usable gain is achieved. The other advantage of this type of circuit is that it also improves the response of the tuned circuit, making it sharper and improving its Q. The 'Q' or quality factor of a tuned circuit is a measure of the performance of a tuned circuit. It is defined as the resonant frequency divided by the bandwidth of the points where the response has fallen by 3 dB.

If the regeneration control is advanced so that the circuit oscillates the receiver is then able to resolve Morse and single sideband signals. The oscillation in the detector beats with a Morse signal to give the characteristic tone. For a single sideband signal it replaces the carrier so that the original audio can be recovered at the detector.

If a regenerative detector is used then it should be preceded by a radio frequency amplifier. This isolates the oscillation generated in the detector from the antenna. If no RF amplification is present then the oscillation can be radiated or transmitted from the antenna and it can cause interference to others nearby.

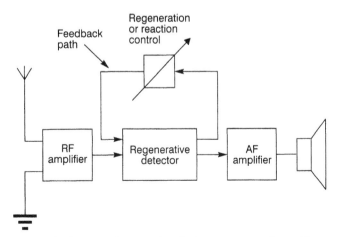

Figure 5.4   *Block diagram of a tuned radio frequency receiver with regeneration*

Direct conversion receiver

One type of radio that has gained popularity over the years and is now widely used as a simple yet effective solution for a variety of applications is the direct conversion set. As its name suggests it uses a process which directly converts the radio frequency signal down to audio frequencies for amplification.

The basic process used in the direct conversion receiver is a process called mixing. This is not like audio mixing where several signals are added together in a linear fashion to give several sounds together. Radio frequency mixing is a non-linear process that involves the level of one signal affecting the level of the other at the output. This process involves the two signal levels multiplying together at any given instant in time and the output is a complex waveform consisting of the product of the two input signals as shown in Figure 5.5. From this it can be imagined that signals on different frequencies are produced, and this is in fact the case.

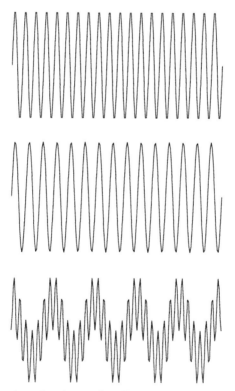

Figure 5.5 *Mixing two signals together. The two top signals are the input signals, and the bottom one is the resulting output*

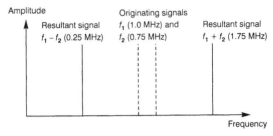

Figure 5.6   *Frequencies produced by mixing two signals together*

It is found that new signals occur at frequencies equal to the sum and difference of the two original signals as shown in Figure 5.6. In other words if the two input frequencies are f_1 and f_2, then the two resultant signals will appear at $f_1 + f_2$ and $f_1 - f_2$. To give an example, if the two original signals are at frequencies of 1 MHz and 0.75 MHz as in Figure 5.5, then the two resultant signals will appear at 1.75 MHz and 0.25 MHz.

If the two frequencies are very near together, for example 1 MHz and 1.001 MHz, then one of the resultant frequencies will appear in the audio part of the spectrum and can be amplified by an audio amplifier and passed into a loudspeaker so that it can be heard.

The circuit used to multiply two signals together is called a mixer or a multiplier as a result of its multiplying action. Either term is equally acceptable, although the term mixer is more commonly used in circuit applications.

To make up a typical direct conversion receiver a number of circuit blocks are required. These are shown in Figure 5.7. In this the signals first pass into a radio frequency amplifier and tuning stage. This fulfils a

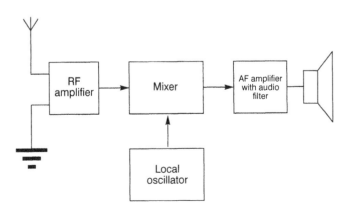

Figure 5.7   *Block diagram of a direct conversion receiver*

number of functions. First of all it amplifies the signals prior to passing them into the mixer. The choice of the level of gain is important, because too much gain can cause the mixer to overload when very strong signals are present. The tuning is also important. If it was not present then signals from a very wide range of frequencies would appear at the input to the mixer and this again might cause the mixer to overload. This is particularly important where very strong signals may appear at frequencies far away from the ones being used. Finally the presence of the radio frequency amplifier helps prevent the local oscillator signal from reaching the antenna. The amplifier acts as another stage of isolation reducing it to an acceptable level.

The local oscillator circuit is crucial to the performance of the radio. It must be capable of tuning over the range that the radio is required to cover and the design and construction of the oscillator is crucial. One of the prime requirements is that it should not drift. If it does then the set will need to be retuned at intervals if the same station is required.

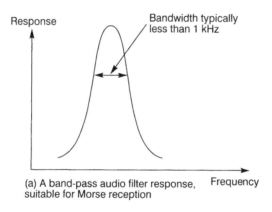

(a) A band-pass audio filter response, suitable for Morse reception

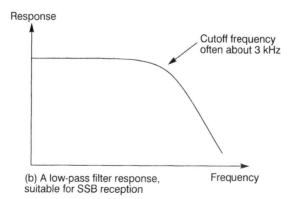

(b) A low-pass filter response, suitable for SSB reception

Figure 5.8 *Band-pass, low-pass filters in a direct conversion receiver*

Once through the mixer the signals enter the audio amplifier. This section may require some filtering. Off-channel signals will produce high frequency interference. This can be reduced by employing an audio filter. For transmissions like Morse a narrow band-pass type of filter with a response like that shown in Figure 5.8(a) can be employed. As the name suggests this will just allow through a narrow band of signals and is particularly useful for removing other signals which may appear in the audio band.

For transmissions like SSB a low-pass filter with a response like that shown in Figure 5.8(b) is generally used. This allows through the low frequency signals and removes those above a certain cutoff frequency. This will help reduce interference from signals off channel which produce high frequency audio interference.

Simple direct conversion receivers can only receive a limited number of types of transmission satisfactorily. Take the example of a Morse transmission. This consists of a carrier wave being turned on and off. When the oscillator in the receiver moves close to the frequency of the Morse signal, an audio note equal in frequency to the difference between the two signals will be generated giving the characteristic on/off tone for Morse. Similarly for single sideband, the oscillator will beat with the sideband to give the required audio. However, for the pitch of the audio to be correct the local oscillator must be on the same frequency that the carrier would have been if it was present.

An amplitude modulated signal is not quite so easy to resolve. If there is a difference in frequency between the carrier of the incoming signal and the local oscillator then an annoying beat note is generated. To be able to resolve the signal satisfactorily the local oscillator must be tuned to exactly the same frequency as the carrier of the required signal. If this is done the audio is recovered and can be heard satisfactorily.

One mode that cannot be resolved easily is frequency modulation. This has meant that this type of receiver is generally only used on the HF bands where Morse and SSB are widely used.

When used for communications applications, one of the problems with a direct conversion receiver is called the audio image and it can be demonstrated as the set is tuned through a steady signal. First, a high frequency audio heterodyne or oscillation is heard. As the receiver is tuned closer to the signal this note falls until eventually the local oscillator of the receiver and the signal are on the same frequency. At this point the heterodyne frequency reaches zero. This point is known as zero beat. As the receiver continues to be tuned in the same direction the audio note starts to rise in frequency as the local oscillator starts to move away on the other side of the signal. This means that there are two points where a particular audio frequency is obtained. This audio image cannot be removed without the addition of more electronics in the mixer. As the requirements for this circuitry are normally quite exacting most direct conversion sets live with the problem.

The direct conversion receiver is finding widespread use in cellular telecommunications applications where its simplicity helps to keep the costs low. Using a sophisticated mixing technique it is possible to extract the transmitted data from the received signal and apply it to the digital circuitry in the phone for processing.

The superhet

To overcome the limitations of receivers such as the TRF or direct conversion set another type of receiver topology evolved and gained popularity in the 1920s and 1930s. Called the superhet, which is short for supersonic heterodyne, this type of set is widely used today. It operates by changing the frequency of the incoming signal to a fixed frequency intermediate stage where it can be more easily filtered and amplified before being demodulated to give an audio signal that is amplified in the normal way.

Using the mixing process it is possible to change the frequency of a signal up or down as shown in Figure 5.9. To illustrate this, take the example of a signal at 2.0 MHz. This mixes with the local oscillator at 1.5 MHz to give a signal at the intermediate frequency of 0.5 MHz or 500 kHz. This signal represents the difference mix product. The sum appears at 3.5 MHz and is easily filtered out as it is very well removed from the frequency of the filters.

Unfortunately there is another mix product which produces a signal that can also pass through the intermediate frequency stages. The first

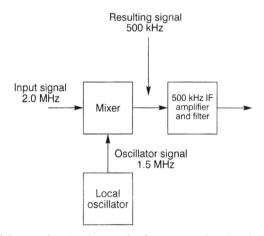

Figure 5.9 *Using a mixer to change the frequency of a signal*

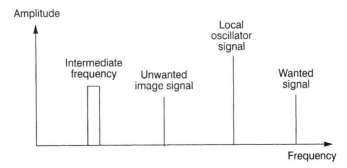

Figure 5.10   *Signals either side of the local oscillator can enter the IF stages*

signal represented the difference between the incoming signal minus the local oscillator. However, a signal representing the local oscillator minus the incoming signal can also pass through the filters. Take the same example again where the local oscillator frequency is 1.5 MHz. An incoming signal of 1.0 MHz also gives an output at 500 kHz. It can be seen that signals which can enter the IF stages are at frequencies equal to the local oscillator frequency plus the intermediate frequency and the local oscillator minus the intermediate frequency as shown in Figure 5.10.

It is obviously undesirable to have two signals on totally different input frequencies which are able to pass through the intermediate frequency filters and amplifiers. Fortunately it is relatively easy to remove the unwanted or image signal to leave only the required one. This is achieved by placing a tuned circuit in the radio frequency stages to remove the signal as shown in Figure 5.11.

From the diagram it can be seen that this tuned circuit does not need to reject signals on adjacent channels like the main intermediate frequency filters. It is only necessary to reject the image signal. This happens to fall

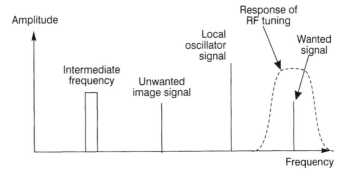

Figure 5.11   *A tuned circuit used to reject the unwanted image signal*

at a frequency equal to twice the intermediate frequency away from the wanted signal. In the case of the example this is 1 MHz away from the wanted signal. This is compared to stations on adjacent channels that are rejected by the main intermediate frequency filters that may be a few kilohertz away, or in some cases less.

Tuning the receiver is accomplished by changing the frequency of the local oscillator. If it is moved up in frequency by 100 kHz then the frequency of the signals which are received will be 100 kHz higher. In the example used earlier, if the local oscillator is 100 kHz higher at 1.6 MHz then signals will be received at 2.1 MHz, assuming the higher frequency signals are wanted and those at 1.1 MHz are rejected as the image.

If the local oscillator and hence the receiver frequency is varied, it is necessary to ensure that the RF tuning also moves at the same rate. In this way the RF tuning will be set to the correct frequency and this will ensure the wanted signal passes through without being attenuated. It also ensures that the image signal is fully rejected.

To ensure that this occurs the tuning for the local oscillator and RF circuits must be linked so that they track together. In many older sets ganged tuning capacitors were used like that shown in Figure 5.12. Here the two sections of the capacitor are mechanically linked. In this way the two tuned circuits can be varied by the same degree at the same time. In more modern receivers electronic methods of tuning are used employing varactor diodes, and these circuits are designed to track at the same rate to maintain the correct circuit conditions.

Figure 5.12 *A ganged tuning capacitor*

Basic superhet receiver

Having looked at the superhet principle the various blocks can be put together to form a complete radio as shown in Figure 5.13. The signals enter the radio frequency circuits from the antenna where the required band of frequencies is selected and those which might give rise to image

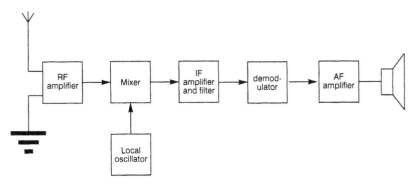

Figure 5.13   *Block diagram of a basic superhet*

Figure 5.14   *A high quality communications receiver that uses the superhet principle (courtesy Icom UK)*

signals are rejected. In the same stages the signals are amplified to the required level and passed into the mixer or multiplier circuit. Here they are mixed with the output from the local oscillator to convert them to the intermediate frequency. As the mixer performs the action of changing the frequency of the signals it is sometimes called a frequency changer.

The signals then pass through the intermediate frequency stages where they are amplified and filtered. The majority of the signal amplification takes place in these stages and the filtering to remove signals on adjacent channels occurs here. In fact the selectivity performance of the whole receiver is dependent upon the filter present in these stages. Accordingly its performance determines that of the whole receiver. As the frequency of the IF stages is fixed, it is possible to design and manufacture highly effective filters.

Finally the signals are demodulated to reconstitute the original signal, and assuming the radio is to be used for audio purposes, the signal is amplified to bring the signal to a suitable level for presenting to the earphones or a loudspeaker.

Overcoming image response problems

The problem of receiving an image signal is one of the major drawbacks of the superhet receiver. Sufficient selectivity needs to be placed into the front end circuits to be able to reduce any unwanted image signals to a sufficiently low level. Unfortunately a receiver covering a wide range of frequencies will have a much greater problem at the top end of the frequency range than at the bottom.

The reason can be deduced from looking at the RF tuned circuits. As the circuit is tuned, the bandwidth increases in proportion with the frequency. As the frequency doubles so does the bandwidth. As the difference between the wanted signal and the image remains constant at twice the intermediate frequency it means that the RF tuning will reject the image signal much less at higher frequencies as shown in Figure 5.15.

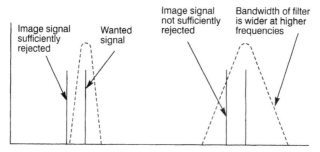

Figure 5.15 *Image response is degraded at higher frequencies*

At higher frequencies the frequency difference becomes a much smaller proportion of the frequency of operation.

There are two main ways of improving the image response. The first is to make the RF tuning much sharper. If this is done then it is necessary to ensure that the RF tuning tracks the local oscillator very accurately. If the improved selectivity of the front end stages is achieved by using more than one tuned circuit, then these have to track one another as well as the local oscillator. Any misalignment of the circuits will cause the wanted signal to be reduced in strength and the image to increase. This can be difficult and for this reason multiple tracked tuned circuits are rarely used.

A more satisfactory option is to increase the intermediate frequency. By doing this the frequency difference between the wanted signal and image is increased. This is the solution that is normally adopted these days because other various techniques are available which allow higher intermediate frequencies to be used.

As image response is an important parameter in performance radio receivers it is a parameter that is often specified, particularly for high performance communications receivers. It is given as a certain number of decibels at a particular frequency. For example, it may be 50 dB at 30 MHz. This indicates that if signals of the same strength on the image and wanted frequencies were present at the input, then the image one would be 50 dB lower than the wanted one at the output of the receiver. The frequency at which the measurement is made has to be included because the figure varies with the frequency of operation. A typical figure for a modern short wave communications receiver would be 80 dB at most frequencies in the short wave spectrum falling to about 60 dB at the worst frequencies.

Multiple conversion sets

The idea of the superhet radio has many advantages over the other types of set which were previously used. It gives better selectivity, it is possible to easily include several types of demodulator which can be switched in as required, it is capable of being made very sensitive, and it can have a whole host of facilities included into its design. However, in its basic form it does have a number of disadvantages. In this first instance its image rejection may be poor, especially at high frequencies as previously described. If the intermediate frequency is increased to improve this then the selectivity may suffer.

To overcome some of these problems multiple conversion receivers may be used. By using more than one conversion the signal may be stepped down to the final IF in two stages. The most obvious method of

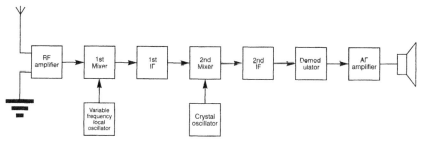

Figure 5.16 *A double conversion superhet with variable frequency first oscillator*

achieving this is to adopt a system as shown in Figure 5.16. Here a variable frequency oscillator is used for the first conversion and a crystal oscillator for the second conversion. In this scheme the first conversion takes the signal down to a higher intermediate frequency to increase the difference between the wanted signal and the image. A second conversion using a fixed frequency oscillator is used to convert the signal down to the second intermediate frequency stage. This oscillator is on a fixed frequency and is usually based around a quartz crystal for convenience and stability.

As the first intermediate frequency is at a higher frequency than before, a major improvement in image response is seen. By keeping the final intermediate frequency stage at a suitably low frequency, then the selectivity performance of the receiver is maintained.

Automatic gain control

A circuit which is standard on all superhet radios these days is called an automatic gain control (AGC). Occasionally this circuit is called an automatic volume control (AVC) although this term is not particularly common these days and the more correct description of AGC is used.

This circuit is used to help compensate for the enormous variation in signal levels that are encountered. From one station to the next there could be a difference of 90 dB or more. Even when the receiver is tuned to one station the signal may still vary. One example of this occurs with car radios where the vehicle is travelling, and hills and other obstructions mean that the signal strength is always varying. Signal strengths on the short wave bands also vary considerably and the AGC can be used to compensate for this.

The AGC operates by sensing the level of the signal when it is demodulated. The voltage generated is used to adjust the gain of some of

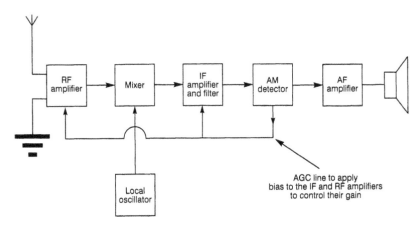

Figure 5.17 *An AGC system in a superhet radio*

the earlier stages. The greater the level of the signal being received, the greater the voltage produced. This enables the gain of the earlier stages to be reduced to a greater extent, thereby keeping the demodulated signal almost constant.

All portable broadcast receivers have AGC circuits that utilize a minimum number of components. However, for more sophisticated receivers like communications sets the design of the AGC system can become considerably more complicated.

The first requirement of the system is that it should not reduce the gain of the set until the signal has achieved a sufficiently high signal to noise ratio. Time constants also have to be taken into consideration. For amplitude modulated signals it is essential that the AGC system does not act so quickly that the modulation is removed from the carrier. To ensure that this does not happen a simple filter is normally used. This may be a simple capacitor resistor network, giving a time constant of about a quarter of a second.

When transmissions such as single sideband are used different time constants are required. When using AM the carrier is present all the time, but this is not so with single sideband. For single sideband the level of the signal varies according to the level of audio present at the transmitter, and it falls to zero when no sound is present. For an AGC system to be able to cope adequately with this type of transmission a dual time constant is needed. It should react very quickly when a new signal appears. In other words it must have a fast attack time. Typically figures of less than 0.1 seconds are used. The decay time must then take account of gaps in the speech or differing levels between syllables in words. Often decay times of a second or more are employed to prevent the AGC level varying unduly and increasing operator fatigue.

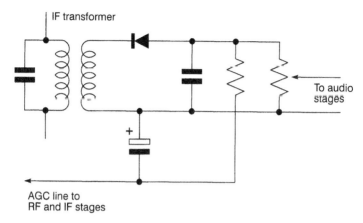

Figure 5.18 *A simple detector giving a voltage output for automatic gain control*

To generate the voltage for the AGC a simple diode detector may be used. One like that shown in Figure 5.18 is often used, especially in medium wave broadcast sets. This performs the normally detection or demodulation function for the audio, but also generates a voltage which is filtered before being applied to the earlier stages of the set to control the gain. In more sophisticated sets a separate detector may be employed. This is likely to have different time constants for each mode being used. In many cases this may be linked to the mode switch on the set, or there may be a separate control on the front panel.

Synthesizers

Many receivers using frequency synthesizers boast such terms as 'PLL', 'Quartz', and 'Synthesized' in their specifications or advertising literature. Frequency synthesizers are found in most of today's receivers. They are able to offer very high degrees of stability, they can operate over a wide range of frequencies, and they easily interface to microprocessor circuitry, enabling them to be controlled by these circuits so that the set can provide new degrees of flexibility and offer a greater number of facilities. Push-button tuning together with frequency memories, scanning and a whole host of other facilities are now available in many sets as a result of the combination of microprocessors and frequency synthesizers.

A frequency synthesizer is based around a phase locked loop which uses the idea of phase comparison for its operation. The phase of a signal is the position within a cycle. Often this is likened to the point on the

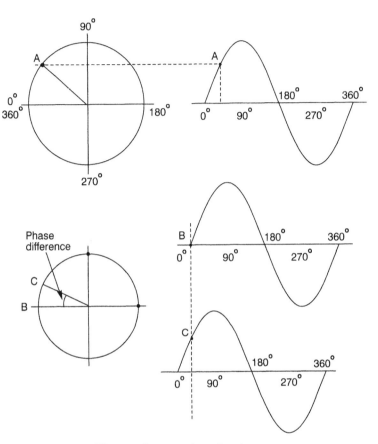

Figure 5.19 *Phase difference between two signals*

signal travelling around a circle with a complete cycle being equal to 360 degrees. Each point on the sine wave has an equivalent point on the circle as shown in Figure 5.19. If there are two signals they may not be at the same point in the cycle and there is a phase difference between them. The phase angle between them is equal to the angular difference on the circle as shown.

From the block diagram of a basic loop shown in Figure 5.20 it can be seen that there are three basic circuit blocks, a phase comparator, voltage controlled oscillator (VCO), and loop filter. A reference oscillator is sometimes included in the block diagram, although this is not strictly part of the basic loop even though a reference signal is required for its operation.

The loop operates by comparing the phase of two signals. The signals from the voltage controlled oscillator and reference enter the phase

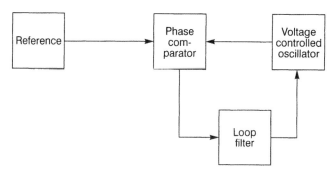

Figure 5.20 *Block diagram of a basic phase locked loop*

comparator and in this circuit a third signal equal to the phase difference between the two is produced. This is passed through the loop filter which performs a number of functions within the loop. In essence it removes any unwanted products from the error voltage before the tune voltage is applied to the control terminal of the voltage controlled oscillator. The error voltage is such that it tries to reduce the error between the two signals entering the phase comparator. This means that the voltage controlled oscillator will be pulled towards the frequency of the reference, and when in lock there is a steady state error voltage. This is proportional to the phase error between the two signals, and is constant. Only when the phase between two signals is changing is there a frequency difference. As there is no phase change when the loop is in lock this means that the frequency of the voltage controlled oscillator is exactly the same as the reference.

A phase locked loop needs some additional circuitry if it is to be converted into a frequency synthesizer. A frequency divider is added into the loop between the voltage controlled oscillator and the phase comparator as shown in Figure 5.21. The divider divides the frequency of

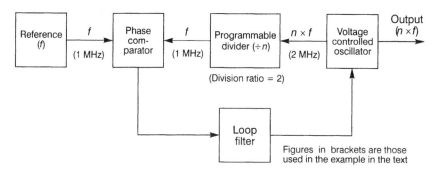

Figure 5.21 *A programmable divider added into a phase locked loop enables the frequency to be changed*

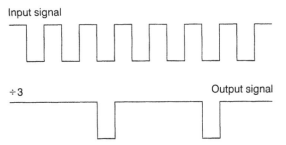

Figure 5.22   *Operation of a divider*

the incoming signal by a certain ratio. If the divide ratio is set to three, then the output frequency is a third of the input, and so forth.

Programmable dividers are widely used in a variety of applications, including many radio frequency uses. Essentially they take in a pulse train like that shown in Figure 5.22, and give out a slower train. In a divide by two circuit only one pulse is given out for every two that are fed in, and so forth. Some are fixed, having only one division ratio. Others are programmable and digital or logic information can be fed into them to set the division ratio.

When the divider is added into the circuit the loop still tries to reduce the phase difference between the two signals entering the phase comparator. Again when the circuit is in lock both signals entering the phase comparator are exactly the same in frequency. For this to be true the voltage controlled oscillator must be running at a frequency equal to the phase comparison frequency times the division ratio as shown in the diagram.

Take the example of when the divider is set to two, and the reference frequency is 1 MHz. When the loop is in lock the frequency entering both ports of the phase comparator will be 1 MHz. If the frequency divider is set to two this means that the frequency entering the divider must be 2 MHz, and the voltage controlled oscillator must be operating at this frequency. If the divider is changed to divide by three, then this means that the voltage controlled oscillator will need to run at 3 MHz, and so forth. From this it can be seen that the loop will increment by a frequency step equal to the comparison frequency when the division ratio is increased by one. In other words the step frequency is equal to the comparison frequency.

As the output frequency from the loop is locked to the reference signal, its stability is also governed by this. Accordingly most reference oscillators use quartz crystals to determine their frequency. In this way any drift is minimized, and the optimum accuracy is obtained.

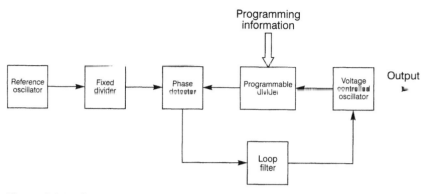

Figure 5.23 *Comparison frequency reduced by adding a fixed divider after the reference oscillator*

Most synthesizers need to be able to step in much smaller increments if they are to be of any use. This means that the comparison frequency must be reduced. This is usually accomplished by running the reference oscillator at a frequency of a megahertz or so, and then dividing this signal down to the required frequency using a fixed divider. In this way a low comparison frequency can be achieved. This gives the block diagram of a basic synthesizer shown in Figure 5.23.

Employing a digital divider is not the only method of making a synthesizer using a phase locked loop. It is also possible to use a mixer in the loop as shown in Figure 5.24. By using a mixer in the loop, an offset to the reference frequency is generated.

The way in which the loop operates with the mixer incorporated can be analysed in the same manner that was used for the loop with a divider.

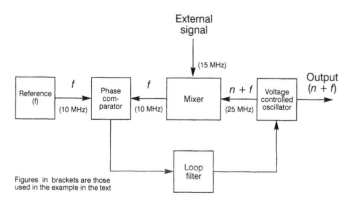

Figure 5.24 *A phase locked loop with mixer*

When the loop is in lock the signals entering the phase detector are at exactly the same frequencies. The mixer adds an offset equal to the frequency of the signal entering the other port of the mixer. To illustrate the way this operates figures have been included. If the reference oscillator is operating at a frequency of 10 MHz and the external signal is at 15 MHz then the VCO must operate at either 5 MHz or 25 MHz. Normally the loop is set up so that the mixer changes the frequency down and if this is the case then the oscillator will be operating at 25 MHz.

It can be seen that there may be problems with the possibility of two mix products being able to give the correct phase comparison frequency. It happens that as a result of the phasing in the loop, only one will enable it to lock. However, to prevent the loop getting into an unwanted state the range of the VCO is limited. For loops which need to operate over a wide range a steering voltage is added to the main tune voltage so that the frequency of the loop is steered into the correct region for required conditions. It is relatively easy to generate a steering voltage by using digital information from a microprocessor and converting this into an analogue voltage using a digital to analogue converter (DAC). The fine tune voltage required to pull the loop into lock is provided by the loop in the normal way.

Multi-loop synthesizers

Many high performance synthesizers use a number of loops, and use both mixers and digital dividers to obtain the required output. By combining a number of loops it is possible to produce high class, wide range oscillators with low levels of phase noise and very small step sizes. If a single loop is used then there are many short falls in the performance.

There are a large variety of ways in which multi-loop synthesizers can be made, dependent upon the requirements of the individual system. However, as an illustration a two loop system is shown in Figure 5.25. This uses one loop to give the smaller steps and the second provides larger steps. This principle can be expanded to give wider ranges and smaller steps.

The operation of the synthesizer is not complicated. The first loop has a digital divider. This operates over the range 19 to 28 MHz and has a reference frequency of 1 MHz to provide steps of 1 MHz. The signal from this loop is fed into the mixer of the second one. The second loop has division ratios of 10 to 19, but as the reference frequency has been divided by 10 to 100 kHz this gives smaller steps.

The operation of the whole loop can be examined by looking at extremes of the frequency range. With the first loop set to its lowest value the divider is set to 19 and the output from the loop is at 19 MHz. This

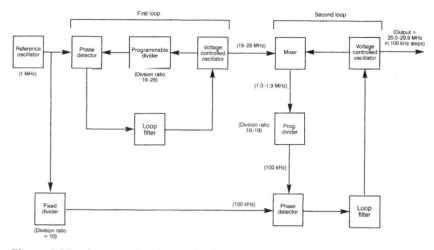

Figure 5.25 *An example of a synthesizer using two loops*

feeds into the second loop. Again this is set to the minimum value and the frequency after the mixer must be at 1.0 MHz. With the input from the first loop at 19 MHz this means that the VCO must operate at 20 MHz if the loop is to remain in lock.

At the other end of the range the divider of the first loop is set to 28, giving a frequency of 28 MHz. The second loop has the divider set to 19, giving a frequency of 1.9 MHz between the mixer and divider. In turn this means that the frequency of the VCO must operate at 29.9 MHz. As the loops can be stepped independently it means that the whole synthesizer can move in steps of 100 kHz between the two extremes of frequency. As mentioned before this principle can be extended to give greater ranges and smaller steps, providing for the needs of modern receivers.

Direct digital synthesizers

Although phase locked loops form the basis of most synthesizers, other techniques can be used. One that is being used increasingly is the direct digital synthesizer (DDS). Although the techniques behind this method of generating signals have been known for many years, it was necessary for semiconductor technology to develop to a sufficient level before they became a viable option. In fact many synthesizers combine both PLL and DDS techniques to gain the best of both systems.

As the name suggests this method generates the waveform directly from digital information. This is in contrast to a phase locked loop system

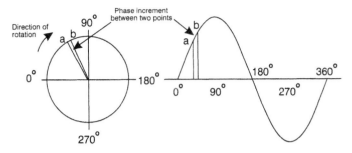

Figure 5.26 *Operation of the phase accumulator in a direct digital synthesizer*

which uses a signal on one frequency, i.e. the reference, to generate a signal on the required frequency by the action of the loop.

The synthesizer operates by storing various points in the waveform in digital form and then recalling them to generate the waveform. Its operation can be explained in more detail by considering the phase advances around a circle as shown in Figure 5.26. As the phase advances around the circle this corresponds to advances in the waveform as shown. The idea of advancing phase is crucial to the operation of the synthesizer as one of the circuit blocks is called a phase accumulator. This is basically a form of counter. When it is clocked it adds a preset number to the one already held. When it fills up, it resets and starts counting from zero again. In other words this corresponds to reaching one complete circle on the phase diagram and restarting again.

Once the phase has been determined it is necessary to convert this into a digital representation of the waveform. This is accomplished using a waveform map. This is a memory which stores a number corresponding to the voltage required for each value of phase on the waveform. In the case of a synthesizer of this nature it is a sine look-up table as a sine wave

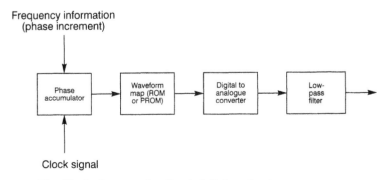

Figure 5.27 *Block diagram of a direct digital synthesizer*

is required. In most cases the memory is either a read only memory (ROM) or programmable read only memory (PROM).

The next stage in the process is to convert the digital numbers into an analogue voltage using a digital to analogue converter (DAC). This signal is filtered to remove any unwanted signals and amplified to give the required level as necessary

There are a number of spurious signals that are generated by a direct digital synthesizer. The most important of these is one called an alias signal. Here images of the signal are generated on either side of the clock frequency and its multiples. For example, if the required signal had a frequency of 3 MHz and the clock was at 10 MHz then alias signals would appear at 7 MHz and 13 MHz as well as 17 MHz and 23 MHz, etc. These can be removed by the use of a low-pass filter.

Tuning is accomplished by varying the size of the step or phase increment between different sample points. By doing this, the speed at which the system advances around the cycle is increased or decreased. This can be done because there will be many more points stored in the waveform map than it is necessary to regenerate the signal. For example, one frequency may be generated by adding 5067 to the value in the phase accumulator each time, whereas the next frequency will be obtained by adding 5068. All the time the clock remains the same, and it is normally a crystal oscillator which may be oven controlled to ensure the frequency stability of the whole system.

From this it can be seen that there is a finite difference between one frequency and the next, and that the minimum frequency difference or frequency resolution is determined by the total number of points available in the phase accumulator. A 24-bit phase accumulator provides just over 16 million points and gives a frequency resolution of about 0.25 Hz when used with a 5 MHz clock. This is more than adequate for most purposes.

It is worth noting that to change frequency the synthesizer simply starts adding the new number onto the total held in the phase accumulator. There is no need for a reset. This means that the synthesizer can change frequency virtually instantaneously. This can be a major advantage of this type of synthesizer in some applications. Phase locked loop systems have a distinct settling time which slows down the rate at which they can change. This may be a critical parameter when receivers are required to scan.

The other advantage of direct digital synthesizers is that they can operate over a very wide range. The major limitation is the top frequency of operation that is governed by the integrated circuits.

These synthesizers do have some disadvantages. They are currently more expensive then their phase locked loop counterparts, but this may change with further developments in semiconductor technology. The other disadvantage is that they produce small spurious signals close to the wanted signal. Normally these can be reduced to acceptable levels, but they may pose a problem in some applications.

Receiver circuitry

Today integrated circuits are widely used in receivers, and indeed there are many ICs that contain complete receivers. However, few IC circuits are shown here as they do not illustrate the principles of the circuits themselves. Instead circuits using discrete components generally lend themselves better to illustrating the principles.

RF amplifier and mixer

The first stages that the signal enters after reaching the input to the receiver are the RF tuning, oscillator and mixer functions. In many portable sets a single transistor performs them all. These circuits are very cost effective to produce, requiring far fewer components than if the circuit was made up by separate circuit blocks. Naturally the performance of the circuit cannot be optimized to give the best performance, but it is normally quite adequate for many of the lower cost sets available on the market.

For higher performance sets high grade mixers are used. There are a number of parameters that are of importance when considering mixer performance. The first is the conversion gain or loss. This is the ratio of the signal level at the RF signal input to the output at the intermediate frequency port. The noise performance is also important as will be described later. As mixers are often located near the front end stages of the set they have a large bearing on the sensitivity and noise performance of the whole set. The isolation between ports is also specified for many ready made mixers. This figure represents the amount of signal from one port that finds its way onto another. The most important are LO to IF and LO to RF input. In a receiver the output frequency is sufficiently different to the local oscillator frequency to enable the leakage signal to be removed by filtering, but in the case of a transmitter this is not always as easy. In the case of the leakage to the RF input, this may cause an unwanted signal to be radiated by the set, and this may cause interference to other nearby users.

A number of circuits can be used as mixers. Diode ring mixers are popular in many high performance applications. Although they give a conversion loss, typically of around 7 dB, other parameters in their performance are normally very high.

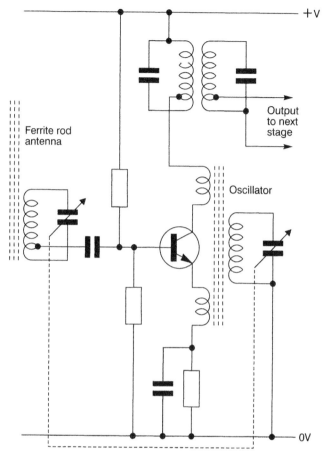

Figure 5.28 *Circuit of a typical self-oscillating mixer stage*

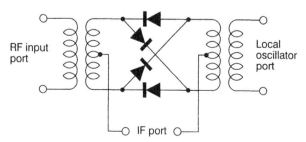

Figure 5.29 *A high performance diode mixer*

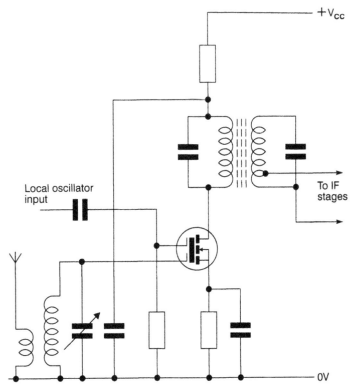

Figure 5.30   *Mixer using a dual gate MOSFET*

Active mixers have a conversion gain. Of these, the circuits based around FETs generally give the best performance because of the characteristics of the devices. They give lower levels of unwanted or spurious signals. Often the circuits use what are called double balanced circuits, but those using a single dual gate MOSFET like that shown in Figure 5.30 perform well and are used in many hi-fi tuners and other radio receivers. In these circuits the signal is applied to one of the gates and the local oscillator or switching signal is applied to the other. The output is taken from the drain or source in the usual way.

Local oscillator

The performance of the local oscillator is very important, particularly where frequency drift is concerned. Today frequency synthesizers are often used. As they normally have a crystal oscillator or crystal oven as their reference, their stability is usually very good. Many older receivers have LC tuned variable frequency oscillators like that shown in Figure 5.31. These oscillators are prone to drift in frequency. This is generally caused by changes in temperature causing the values of some of the components to alter slightly. Even very small changes will give noticeable amounts of drift. If the oscillator drifts then the set will need to be retuned. This is not usually a problem from long and medium wave sets, but where frequencies are higher and modes of transmission like SSB are used the stability of the oscillator is of great importance.

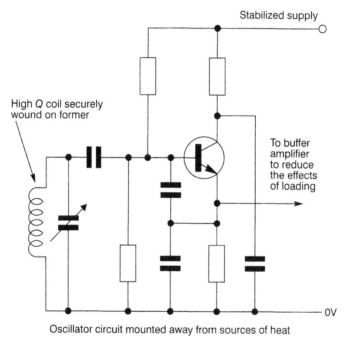

Oscillator circuit mounted away from sources of heat

Figure 5.31 *An LC tuned variable frequency oscillator*

To overcome the effects of drift a number of precautions must be employed. The circuit should be run from a stabilized power supply. Changes in supply voltage cause the circuit parameters of some components to change. This is particularly true of some semiconductor devices like bipolar transistors. Coils should be wound as securely as possible and the tuned circuit should not be switched if at all possible. In addition to this the tuned circuit should be designed to have a high *Q*. Finally the whole oscillator circuit must be kept away from any sources of heat.

IF amplifier and filter

After the mixer, the signals reach the IF stages. It is here that the majority of the gain and selectivity are provided. Most of today's sets utilize integrated circuits as the basic building blocks, and the actual circuits depend upon the particular IC being used. Some radios still use discrete components and a typical circuit is shown in Figure 5.32. Often two stages are used in series to give the required gain and selectivity. In this circuit the selectivity is provided by the coupling transformer. This is tuned using a ferrite adjuster and each one requires adjustment to the correct frequency.

A number of standard intermediate frequencies are used. Frequencies of 455 or 465 kHz are widely used for medium and long wave broadcast radios. The low frequency means that it is possible to obtain sufficient selectivity with a minimum number of stages.

Other frequencies are used for other purposes. A frequency of 10.7 MHz is a standard for VHF FM receivers, and this frequency is also used in many hand-held transceivers for mobile or on-site communications. Frequencies of 1.6 and 9 MHz are also used in some high performance short wave communications receivers.

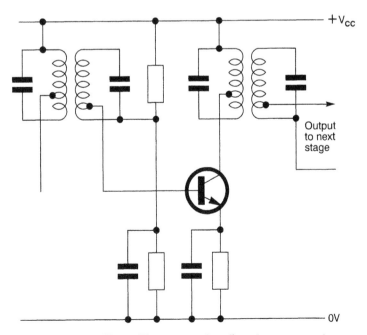

Figure 5.32   *A typical IF amplifier stage using discrete components*

The main filtering for the receiver is undertaken in the IF stages. It provides the selectivity that accepts stations on the required channel, and rejects those on the next channel. The filtering can take many forms dependent upon the receiver circuitry. One of the most basic is based around LC tuned circuits. They consist of inductors and capacitors and are often in the form of tuned transformers used to couple the individual stages together as shown in Figure 5.32. In many older style portable sets used for broadcast reception two or three of these transformers are used to give the required degree of selectivity, although they do not provide the required levels of performance for many other applications.

A further disadvantage of LC filters is that they require alignment. If transformers are used then each one has its own ferrite tuning core. This is required so that the resonant frequency of the circuit can be tuned to exactly what is required. These have to be tuned to the correct frequency during the production of the set. Normally each tuned circuit is slightly offset from the others so that the correct bandwidth is achieved. If they were all placed on exactly the same frequency then the bandwidth would be too narrow.

When better performance is required there are a number of options that can be taken. One is to use a filter based around one or more quartz crystals. These crystals have very high levels of *Q* and accordingly they are able to make exceptionally selective filters. Being more expensive than simple LC tuned circuits they are not used in low cost broadcast sets, but they are often found in more expensive specialist receivers where very high levels of performance are required. The operation of quartz crystals depends upon the piezoelectric effect. This links the electrical circuit to the very high *Q* mechanical resonances that these crystals exhibit. As a result quartz crystals possess *Q*s of many thousands and they are able to provide very high levels of selectivity.

A single crystal on its own does not give an ideal response curve. The response is not symmetrical, and has a very narrow peak. This is not ideal for all types of transmission as they need a certain bandwidth to accept all of the signal. To obtain a much better performance in terms of the ultimate rejection and required bandwidth, a number of crystals are used together. One popular circuit is called a half lattice filter, and this is shown in Figure 5.33. The response curve for this is far more even as shown in the diagram, although there is some ripple. To achieve the wider bandwidth the resonant frequencies of the two crystals are spaced apart slightly. As a rough guide the bandwidth of the filter at the −3 dB points is approximately 1.5 times the difference in the resonant frequencies of the crystals.

Two important factors of any filter are the rate at which the response falls away and the ultimate rejection. These can both be improved by adding further stages to the filter. A two pole filter, i.e. one with two

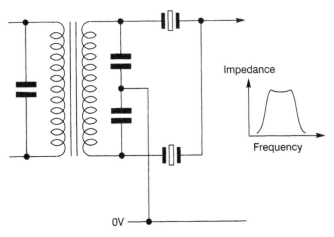

Figure 5.33 *A half lattice crystal filter circuit*

crystals like that shown in the diagram, is rarely sufficient to meet the requirements in a receiver. Accordingly six or eight pole versions are more common.

Quartz crystals are not the only way in which filters with a better performance than LC filters can be made. A variety of different substances exhibit the piezoelectric effect. A number of ceramics exhibit it, although it is not combined with such a sharp mechanical resonance. Even so they give a much better performance than their LC counterparts.

Ceramic filters operate in exactly the same way as quartz filters, the signal being linked to the mechanical resonances in the substance. As the resonance is less sharp, the bandwidths of these filters is wider and the shape factor is poorer. Filters made from these ceramics are much cheaper to manufacture than quartz ones, and as a result they find widespread use in many areas. Many millions are used each year in broadcast receivers where they are ideal in terms of cost and performance. They are never used in performance sets as the prime source of filtering, although they are sometimes included where a small degree of filtering is used. However, as the performance of these filters is being improved all the time, they are likely to be found increasingly.

Ceramic filters are available with centre frequencies that range from a few kilohertz up to ten megahertz and more and they possess bandwidths ranging from around 0.05 per cent up to 20 per cent. A simple ceramic resonator consists of a disc with electrodes plated on opposite faces. These items perform in a similar way to a single quartz crystal. More commonly used is a resonator with one of the electrodes separated to form two isolated parts. This forms a three electrode network that can be incorporated into a circuit such as that shown in

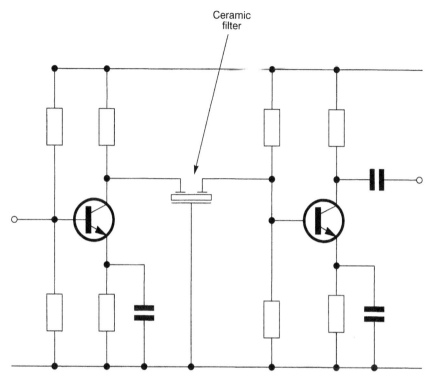

Figure 5.34 *A circuit using a ceramic filter*

Figure 5.34. They are widely used with integrated circuit IF strips where they are the only filter providing the selectivity.

Demodulation

The signals that appear at the output of the intermediate frequency amplifier and filter section of a receiver consist of a radio frequency type of signal. In other words they consist of a carrier with the required modulation superimposed. To remove the modulation, a circuit called a demodulator is required. While these circuits are often termed demodulators, the name detector is also used almost interchangeably to describe them as well.

The type of demodulator and hence the circuitry of the demodulator will depend upon the type of modulation which is being used. Often different demodulators are required for different modes of transmission. When a receiver needs to be able to demodulate a variety of types of signal, different detectors are switched in.

AM demodulation

The simplest type of detector is a simple diode detector like that shown in Figure 3.4. This type of detector is widely used in many receivers, although it can give relatively high levels of distortion for some applications. It is also not particularly good when selective fading occurs on short wave signals. Under these conditions certain of the frequencies of a signal are not received for short periods of time, and this can lead to very high levels of distortion.

SSB and Morse demodulation

To receive Morse and single sideband it is necessary to have additional circuitry. In order to hear the Morse signal it is necessary to beat the incoming signal with an internally generated one so that the typical Morse signal of a tone being turned on and off can be heard. If a normal AM detector was used, then a series of clicks and thumps would be heard as the carrier was turned on and off.

The internal oscillator is generally called a beat frequency oscillator or BFO. The incoming signal and the beat signal are passed into a mixer where they are mixed together. The resulting audio tone is equal in frequency to the difference between the signal and the BFO. As the mixing process is a multiplication process, and the output is the product of the two input signals, this type of detector is often called a product detector.

The same type of detector is also used for receiving single sideband. Here the beat frequency oscillator is used to reintroduce the carrier as shown in Figure 5.35. As a result of this application the oscillator may also be known as the carrier insertion oscillator or CIO.

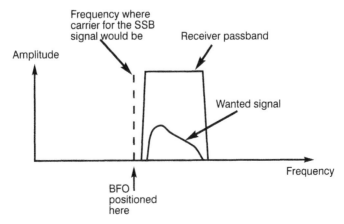

Figure 5.35   *Action of a beat frequency oscillator*

The frequency of the BFO needs to be set to give the optimum performance. The signal being received should be tuned for the best reception, by placing it in the middle of the receiver pass-band. The setting of the oscillator is then adjusted to suit. This may often mean that it is just outside the pass-band of the set. However, this does not matter as it comes after the selectivity

Synchronous AM demodulation

Improvements in amplitude modulation reception can be achieved by using a system known as synchronous detection. Essentially it uses a beat frequency oscillator and product detector or mixer for AM. For this to work the beat frequency oscillator must be kept on the same frequency or zero beat with the carrier. There are a number of methods of doing this, but the one which is most commonly used employs a high gain limiting amplifier. This takes the AM signal and removes the modulation to leave only the carrier as shown in Figure 5.36. This is used as the beat frequency oscillation to demodulate the signal.

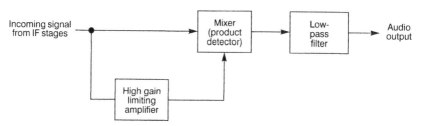

Figure 5.36 *Synchronous detector*

Synchronous detection of AM gives a far more linear method of demodulation. It also gives more immunity against the effects of selective fading on the short wave bands.

FM demodulation

In order to be able to demodulate an FM signal it is necessary to convert the frequency variations into voltage variations. In order to be able to achieve this the circuit of the demodulator needs to be frequency dependent. The ideal response is a perfectly linear voltage to frequency characteristic like that shown in Figure 5.37. Here it can be seen that the centre frequency is in the middle of the response curve and this is where

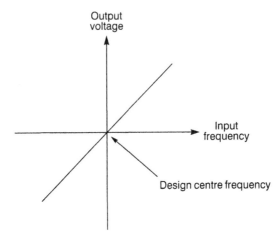

Figure 5.37   *An ideal FM demodulator response curve*

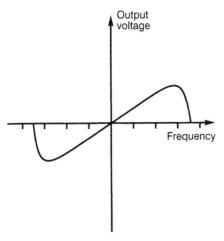

Figure 5.38   *The typical characteristic curve of a real FM demodulator*

the unmodulated carrier would be located when the receiver is correctly tuned into the signal. In other words there would be no offset DC voltage present. In most cases the demodulator has a response like that shown in Figure 5.38. For obvious reasons the response curve is known as an 'S' curve. From this it can be seen that as the signal moves up in frequency a higher voltage is produced and vice versa. In this way the frequency variations of the signal are converted into voltage variations which can be amplified by an audio amplifier before being passed into headphones or a loudspeaker.

To enable the best detection to take place the signal should be centred about the middle of the curve. If it moves off too far then the characteristic becomes less linear and higher levels of distortion result. Often the linear region is designed to extend well beyond the bandwidth of a signal so that this does not occur. In this way the optimum linearity is achieved. Typically the bandwidth of a circuit for receiving VHF FM broadcasts may be about 1 MHz whereas the signal is only 200 kHz wide.

In many receivers it is necessary to ensure that the receiver stays on the correct frequency. Any drift in the set may mean that the signal may not remain on the linear portion of the demodulator curve. Some sets incorporate a circuit called an automatic frequency control or AFC. This enables the set to remain on tune, even when the local oscillator is prone to drifting. The circuit operates by taking the output from the demodulator and removing the audio using a simple filter. The remaining voltage is then an indication of where the carrier is on the demodulator curve. Ideally it should be in the centre. The voltage can be applied to the local oscillator in the set as a correction signal to pull its frequency slightly so that the signal remains in the centre of the pass-band.

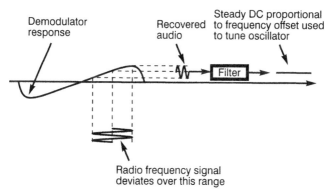

Figure 5.39 *Operation of an AFC*

One of the advantages of FM is its resilience to noise. This is one of the main reasons why it is used for high quality audio broadcasts. However, when no signal is present, a high noise level is present at the output of the receiver. If a low level FM signal is introduced and its level slowly increased it will be found that the noise level reduces. From this the quieting level can be deduced. It is the reduction in noise level expressed in decibels when a signal of a given strength is introduced to the input of the set. Typically a broadcast tuner may give a quieting level of 30 dB for an input level of just over a microvolt.

To remove the high level of noise produced when no signal is present, a 'squelch' circuit may be included. This mutes the audio when no signal is present and there is a high level of noise at the output. Many scanners and hand-held VHF or UHF transceivers have a control to adjust the signal level below which the audio cuts off. By using this control the set can be adjusted to detect very low level signals if necessary. Hi-fi tuners also incorporate a squelch circuit in most instances. However, they seldom have a control to adjust as it is usually not a requirement to hear signals which are only just above the noise level.

Another effect which is often associated with FM is called the capture effect. This can be demonstrated when two signals are present on the same frequency. In cases of this nature, only the stronger signal will be heard at the output, unlike AM where a mixture of the two signals is heard, along with a heterodyne if there is a frequency difference.

The capture effect is often defined in receiver specifications as the capture ratio. It is defined as the ratio between the wanted and unwanted signal to give a certain level of the unwanted signal. Normally a reduction of the unwanted signal of 30 dB is used. To give an example of this the capture ratio may be 2 dB for a typical tuner to give a reduction of 30 dB in the unwanted signal. In other words if the wanted signal is only 2 dB stronger than the unwanted one, the audio level of the unwanted one will be suppressed by 30 dB.

A number of circuits can be used to demodulate FM. The very simplest form of FM demodulation is known as slope detection or demodulation. It consists of a tuned circuit that is tuned to a frequency slightly offset

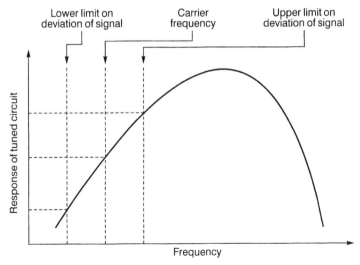

Figure 5.40 *Concept of a slope detector*

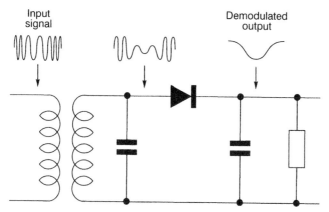

Figure 5.41 *Stages of demodulation when using a slope detector or demodulator*

from the carrier of the signal. As the frequency of the signals varies up and down in frequency according to its modulation, so the signal moves up and down the slope of the tuned circuit. This causes the amplitude of the signal to vary in line with the frequency variations. In fact at this point the signal has both frequency and amplitude variations. The final stage in the process is to demodulate the amplitude modulation and this can be achieved using a simple diode circuit. One of the most obvious disadvantages of this simple approach is the fact that both amplitude and frequency variations in the incoming signal appear at the output. However, the amplitude variations can be removed by placing a limiter before the detector. Additionally the circuit is not particularly efficient as it operates down the slope of the tuned circuit. It is also unlikely to be particularly linear, especially if it is operated close to the resonant point to minimize the signal loss.

For many years circuits employing discrete components were used and the two most popular circuits were the ratio detector or demodulator and the Foster-Seeley demodulator. Of these the ratio detector was the most popular as it offered a better level of rejection of amplitude modulation, and hence it would operate satisfactorily with poorer levels of limiting in the preceding IF stages of the receiver.

The circuit of the ratio detector is shown in Figure 5.42 and essentially operates by using a frequency sensitive phase shift. It is characterized by the special transformer and the diodes that are effectively in series with one another. When a steady carrier is applied to the circuit the diodes act to produce a steady voltage across the resistors R_1 and R_2, and the capacitor C_3 charges up as a result.

The transformer enables the circuit to detect changes in the frequency of the incoming signal. It has three windings. The primary and secondary act in the normal way to produce a signal at the output. The third winding is untuned and the coupling between the primary and the third winding is very tight, and this means that the phasing between signals in these two windings is the same.

The primary and secondary windings are tuned and lightly coupled. This means that there is a phase difference of 90 degrees between the signals in these windings at the centre frequency. If the signal moves away from the centre frequency the phase difference will change. In turn the phase difference between the secondary and third windings also varies. When this occurs the voltage will subtract from one side of the secondary and add to the other causing an imbalance across the resistors R_1 and R_2. As a result this causes a current to flow in the third winding and the modulation to appear at the output.

The capacitors C_1 and C_2 filter any remaining RF signal which may appear across the resistors. The capacitor C_4 and R_3 also act as filters ensuring no RF reaches the audio section of the receiver.

The Foster-Seeley detector is very similar to the ratio detector in many ways, but there are a few fundamental differences. Although the circuit

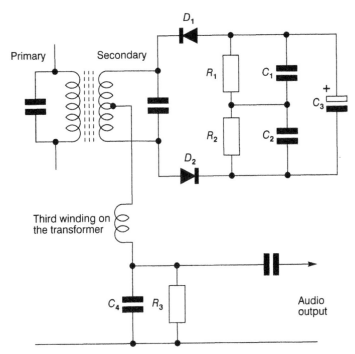

Figure 5.42 *Circuit of the ratio detector*

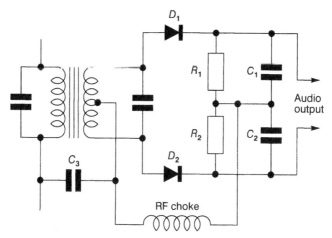

Figure 5.43 *Circuit of the Foster-Seeley detector*

topology looks very similar, having a transformer and a pair of diodes, there is no third winding. Instead this is replaced with a choke.

While there are a number of differences the circuit again operates on the phase difference between signals. To obtain the different phased signals a connection is made to the primary side of the transformer using a capacitor, and this is taken to the centre tap of the transformer. This gives a signal that is 90 degrees out of phase.

When an unmodulated carrier is present at the centre frequency, both diodes conduct, producing voltages across their respective load resistors that are equal and opposite. These two voltages cancel one another out at the output so that no voltage is present. As the carrier moves off to one side of the centre frequency the balance condition is destroyed, and one diode will conduct more than the other. This results in the voltage across one of the resistors being larger than the other, and a resulting voltage at the output corresponding to the modulation on the incoming signal.

The choke is required in the circuit to ensure that no RF signals appear at the output. The capacitors C_1 and C_2 provide a similar filtering function.

One of the major problems with both of these circuits is that they require coils and chokes. These are expensive to produce these days, often costing more than an integrated circuit. As a result other methods of demodulating FM are often used.

A third form of demodulator is known as a quadrature of coincidence detector. It is widely used in integrated circuits and provides a good level of linearity. It has the advantage over systems such as the ratio and Foster-Seeley detectors in that it requires a simple tuned circuit rather

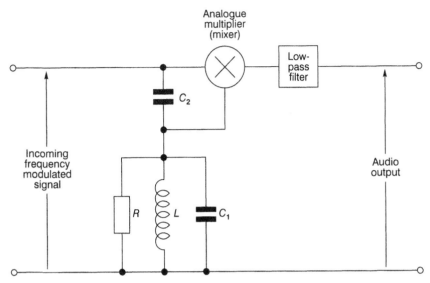

Figure 5.44   *Block diagram of an FM quadrature detector*

than the more complicated transformers they use. In this way costs can be reduced. It is also very easy to implement in a form that is applicable to integrated circuits.

The basic format of the quadrature detector is shown in Figure 5.44. It can be seen that the signal is split into two components. One of these passes through a network that provides a basic 90 degree phase shift, plus an element of phase shift dependent upon the deviation. The original signal and the phase shifted signal are then passed into a multiplier or mixer. It is found that the mixer output is dependent upon the phase difference between the two signals, i.e. it acts as a phase detector and produces a voltage output that is proportional to the phase difference and hence to the level of deviation on the signal.

The detector is able to operate with relatively low input levels, typically down to levels of around 100 microvolts and it is very easy to set up requiring only the phase shift network to be tuned to the centre frequency of the expected signal. It also provides good linearity and this results in low levels of distortion.

Often the analogue multiplier is replaced by a logic AND gate and the input signal is hard limited to produce a variable frequency pulse waveform. The operation of the circuit is fundamentally the same, but it is known as a coincidence detector. Also the output of the AND gate has an integrator to 'average' the output waveform to provide the required audio output, otherwise it would consist of a series of square wave pulses.

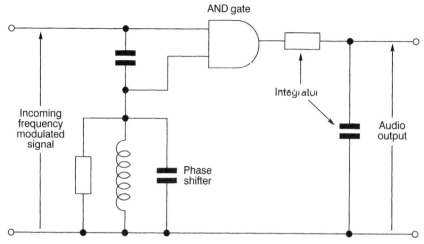

Figure 5.45 *A coincidence detector*

Phase locked loops are widely used for demodulating FM. Apart from not using a coil they also give a very linear voltage to frequency conversion. As a result of this they are often used in hi-fi tuners.

The way in which they operate is very simple. The circuit is set up to operate as shown in Figure 5.46. The FM signal from the IF stages of the set is connected to one of the phase detector inputs as shown, and the output from the VCO is connected to the other.

With no modulation applied and the carrier in the centre position of the pass-band the voltage on the tune line to the VCO is set to the mid position. However, if the carrier deviates in frequency, the loop will try to keep the loop in lock. For this to happen the VCO frequency must follow the incoming signal, and for this to occur the tune line voltage must vary. Monitoring the tune line shows that the variations in voltage correspond to the modulation applied to the signal. By amplifying the variations in

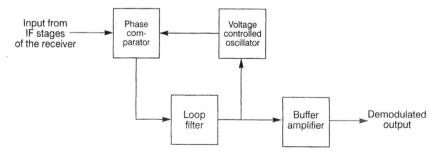

Figure 5.46 *A phase locked loop FM demodulator*

voltage on the tune line it is possible to generate the demodulated signal.

It is found that the linearity of this type of detector is governed by the voltage to frequency characteristic of the VCO. As it normally only swings over a small portion of its bandwidth, and the characteristic can be made relatively linear, the distortion levels from phase locked loop demodulators are normally very low. Many tuners have specifications of fractions of a per cent of distortion.

Digital signal processing (DSP)

Microprocessors are finding uses in an increasing number of applications. It is therefore hardly surprising to find that they are being used in a variety of areas that normally used only analogue electronics. Functions including filtering, mixing, demodulation and the like can all be performed digitally by specialized digital signal processors. As a result many receivers use these techniques, and in the coming years this process will become far more widespread. In fact digital radio (DAB) receivers rely very heavily on these techniques.

The process is based upon the fact that it is possible to build up a representation of the signal in a digital form. This is done by sampling the voltage level at regular time intervals and converting the voltage level at that instant into a digital number proportional to the voltage as shown in Figure 5.47. This process is performed by a circuit called an analogue to digital converter, A to D converter or ADC. In order that the ADC is presented with a steady voltage while it is taking its sample, a sample and hold circuit is used to sample the voltage just prior to the conversion. Once complete the sample and hold circuit is ready to update the voltage

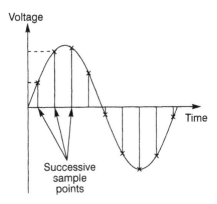

Figure 5.47 *Sampling a waveform*

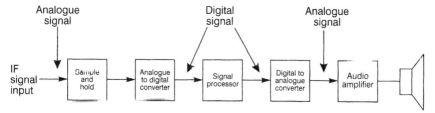

Figure 5.48 *Block diagram of a digital signal processor*

again ready for the next conversion. In this way a succession of samples is made.

Once in a digital format the processor performs complicated mathematical routines upon the representation of the signal. However, to use the signal it then needs to be converted back into an analogue form where it can be amplified and passed into a loudspeaker or headphones as shown in Figure 5.48. The circuit that performs this function is not surprisingly called a digital to analogue converter, D to A converter or DAC.

The advantage of digital signal processing is that once the signals are converted into a digital format they can be manipulated mathematically. This gives the advantage that all the signals can be treated far more exactly, and this enables better filtering, demodulation and general manipulation of the signal. Unfortunately it does not mean that filters can be made with infinitely steep sides because there are mathematical limitations to what can be accomplished.

IF breakthrough

One problem which can occur happens when signals from the antenna pass across the RF sections and enter the IF directly. Normally intermediate frequencies are chosen to fall at frequencies which are not used by high powered stations. Although this alleviates the problem to a large extent, there are sometimes situations where this is not feasible.

To ensure that breakthrough is not a problem care must be taken during the design of the receiver to ensure that there is sufficient isolation between the signal input and the IF stages. In some sets special filters are included at the antenna to ensure that signals from the antenna at the intermediate frequency are removed as they enter the set.

Figures for IF breakthrough are quoted in the same way as they are from image rejection. In the case of most high performance receivers it is possible to achieve figures of more than 60 dB rejection. In some cases figures of 100 dB have been quoted.

Spurious signals

Virtually all receivers generate signals themselves, many of which can by picked up be the set and received as if they had entered the set through the antenna socket. In many cases these signals go unnoticed because they are masked out by real signals, but in some instances they act as unwanted interference degrading the performance of the set.

Spurious signals can be generated in a number of ways. Most of today's receivers use a variety of oscillators in their circuitry. It is very easy for these signals, or their harmonics to enter the signal path and appear as signals.

On wideband receivers it is often impossible to eliminate all the spurious signals, although most of them can be reduced to levels where they do not cause a major problem. Some receivers will quote known spurious signals.

For narrow-band receivers like those in mobile phones it is possible to design the set so that no significant spurious signals are generated within the frequency range of the receiver.

Phase noise and reciprocal mixing

One of the main problems with frequency synthesizers is the fact that they can generate high levels of phase noise if care is not taken in the design. This noise is caused by small amounts of phase jitter on the signal, and it manifests itself as noise spreading out either side of the signal as shown in Figure 5.49.

Any signal source will have some phase noise. Crystal oscillators are very good, and free running variable frequency oscillators normally perform well. Unfortunately synthesizers, especially those based around phase locked loops, do not always fare so well and this can adversely affect the performance of the radio in terms of reciprocal mixing.

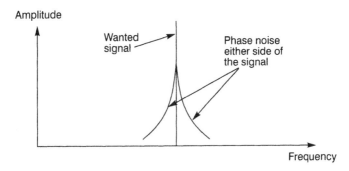

Figure 5.49   *Phase noise on a signal*

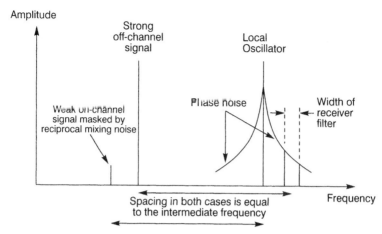

Figure 5.50 *Reciprocal mixing*

This can be explained by considering the receiver when it is tuned to a strong signal. The local oscillator mixes with the wanted station to produce phase noise which falls within the pass-band of the receiver filters. If the receiver is tuned off channel by a given amount, for example 10 kilohertz, then the station can mix with the local oscillator phase noise 10 kilohertz away from the main oscillator signal as shown in Figure 5.50. If the incoming station signal is very strong the reciprocal mixing effect can mask out weaker stations, thereby reducing the effective sensitivity of the set in the presence of strong off-channel signals.

Filters

There is a wide variety of different types of filter which are found in today's receivers. LC, crystal, ceramic, audio filters and more can all be found. Each one has its own advantages and disadvantages, and by careful use of the correct type in particular positions in the set, excellent selectivity can be obtained.

The basic selectivity of a set is provided in the IF stages. Here the stations on adjacent channels are rejected, and the performance of the filters here will determine the performance of the whole set. Often several filters placed at different points in the set provide the selectivity. However, in most sets one high quality unit is used to provide the majority of the selectivity. In multi-conversion sets this filter is placed at the last intermediate frequency where the frequency is generally the lowest. In this way it is possible to achieve the most effective filtering.

Selectivity and filters

Each filter has a finite bandwidth, and this must be taken into consideration when selecting a filter for a receiver. As transmissions have a finite bandwidth which depends upon the type of transmission in use it is necessary to match the filter bandwidth to the transmission. If the filter is too wide as shown in Figure 5.51, then there is the possibility of

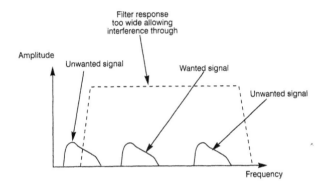

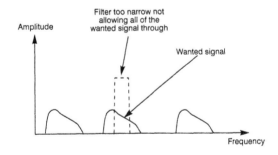

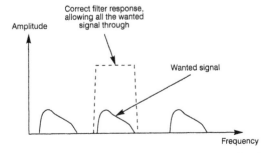

Figure 5.51 *Filter bandwidths must match the transmission bandwidth*

picking up unwanted signals. On the other hand, if it is too narrow then part of the wanted signal will be rejected and this will result in distortion. It is therefore necessary to use filters of the required bandwidth. Typically medium wave AM transmissions need about 9 or 10 kHz, and the standard for short wave AM transmissions is about 6 kHz while for SSB a bandwidth of 2.7 kHz or thereabouts is usually used. The VHF FM broadcasts normally require a bandwidth of about 200 kHz to ensure the whole of the signal is received, and other types of transmission all have their own requirements. Other types of transmission will occupy different bandwidths and filters will be required to accommodate the required bandwidth.

As the bandwidth of the filter in the IF stages of the receiver determines the close in selectivity of the receiver, it is necessary to be able to specify its performance accurately. One of the main aspects of the performance of a filter is its pass-band. This is the band of frequencies that the filter allows through. In an ideal world, it would allow through signals within the pass-band, and totally reject all others that fall outside the pass-band as shown in Figure 5.52.

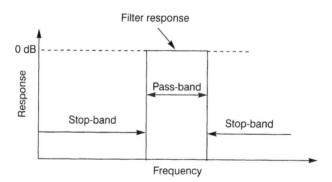

Figure 5.52 *An ideal filter response*

The response of a real filter like one used in a receiver to give it its selectivity is different and is shown in Figure 5.53. From this it can be seen that the response curve does not change instantly from accepting signals to rejecting them. Also it does not have an infinite rejection of unwanted signals. Nevertheless many filters achieve very good results.

The first point which needs to be noted on any filter is its pass-band. This is the bandwidth over which the filter accepts signals. The limits of the pass-band are normally taken to be the points where the response has fallen by 6 dB. For a filter used in a communications receiver to receive an AM signal on the short wave bands the bandwidth may be 6 kHz.

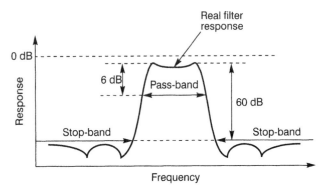

Figure 5.53   *A typical filter response*

The stop-band is also of interest. For this the bandwidth for a given attenuation is taken. The normal value for this is 60 dB although again other values may be taken. As a result the bandwidth and attenuation levels should be specified. Typically a filter might be specified as having a stop-band bandwidth of 6 kHz at −60 dB.

The rate at which the filter reaches its final attenuation is of interest. Even though it may have the correct pass-band, if it does not reach its final attenuation fast enough, it will let interference through. The shape factor is a measure of the rate at which it reaches its stop-band. This is taken as the stop-band bandwidth divided by the pass-band bandwidth. Again the levels of attenuation must be given. Taking the figures already used as an example the shape factor would be 2.2:1 at 6/60 dB.

It is also noticed that the filter response is not flat. From the diagram it can be seen that there is ripple within the pass-band, and also in the stop-band. Pass-band ripple is generally only a few decibels and this is often quoted in the specification for the filter. Typical values for the filter may be one or two decibels. The stop-band ripple is also of importance and can easily be tens of decibels.

Filter also introduces a certain amount of loss. This can be important for circuit designers because the effect of any losses has to be taken into consideration when calculating signal levels at certain points in the circuit. It may be in some circumstances that there will be a balance between the degree of rejection of unwanted signals and the loss of the filter. Generally the greater the rejection of unwanted signals, the more stages are required in the filter, and the greater the in-band loss.

The final rejection of the unwanted signals is another important parameter. Any filter will only be able to reject signals outside the pass-band by a certain amount. Normally the figure for the final rejection is taken so that any ripple in the stop-band does not bring the response above the figure for the final rejection. When designing and constructing a filter it is necessary to ensure that there are not any leakage paths between the input and the output of the filter. Sometimes there may be a small amount of capacitance between the two connections and this can allow signals to leak across the filter, thereby reducing the level of the final rejection.

Sensitivity and noise

One of the most important aspects of any radio receiver is its sensitivity, or ability to pick up weak signals. Most receivers today are very sensitive and able to pick up signals which are less than a microvolt at the input to the set. Unfortunately a set cannot be made more sensitive by simply adding more stages of amplification as a number of other aspects of receiver design quickly become obvious and limit the sensitivity. The first is that if there is too much gain then some of the stages in the set become overloaded leading to other problems which will be described later. The other problem is that of noise and this is the major limiting factor. There is a certain amount of noise which is present at the input to the receiver, and each stage within the set adds a little more.

Noise consists of random electrical impulses and is present at all frequencies, although the levels and sources vary with frequency. In many cases it is 'white noise' which is heard as a background hiss and it is present at all frequencies. Once it is present it cannot be removed or cancelled out. It can only be reduced by limiting the bandwidth of the set, and if the bandwidth is reduced too much then the wanted signal may be impaired.

Other types of noise may appear as bangs or pops, and this type of noise is heard particularly on the lower frequency bands.

This external noise comes from two main areas. It may be picked up by the antenna. In turn this can come from a variety of sources. It may be man-made. The variety of electrical and electronic equipment in use today generates noise which can be picked up over the whole of the radio spectrum. Electric motors, vehicle ignition systems, televisions, computers and even fluorescent lights all produce energy which can be picked up.

Naturally occurring noise can also be picked up. Again there are a variety of sources for this. Static is one major type of noise. This comes from discharges in electrical storms. In view of the colossal amounts of energy involved in a static discharge it is hardly surprising it generates large amounts of radio frequency energy which can be heard over large areas. Noise also comes from outer space. This cosmic or galactic noise comes from many places in outer space, from our own sun, to the distant galaxies. Surprisingly this noise can be present at significant levels, despite the enormous distances involved.

The level of different types of noise varies with frequency, and this means that the design of radios is dependent to a certain degree upon this. Some forms of man-made noise extend to relatively high frequencies. Car ignition noise can extend to frequencies in excess of 400 MHz and noise from fluorescent lights can also be picked at frequencies well above 1 GHz.

Naturally occurring forms of noise fall with increasing frequency. Atmospheric noise predominates at low frequencies, but above about 10 MHz galactic noise is the main constituent. Even this falls with increasing frequency leaving that generated within the receiver as the main source of noise at frequencies above about 50 MHz.

Receiver noise performance is fashioned to a large degree by the levels of the different types of noise which occur. In view of the high levels of noise entering a receiver from the antenna on frequencies below 30 MHz there is little point in designing very low noise receivers. At frequencies above about 50 MHz to 100 MHz received noise levels fall to a level where the noise generated within the receiver predominates, and accordingly the noise performance of the set becomes far more important.

Another form of noise is called thermal noise. This is caused by the movement of the electrons in a conductor. Even when there is no potential difference to cause current to flow noise is present. At any temperature above absolute zero electrons are moving about randomly in the conductor, and this causes random voltages or noise to be generated. It is not surprising that the level of noise is proportional to the temperature. As the temperature rises so the movement of the electrons increases. It is also found that the level of noise is proportional to the bandwidth being received. The wider the bandwidth, the greater number of noise frequencies which can be received, and hence the greater the level of noise. Finally the noise is proportional to the resistance of the conductor. In fact the level of noise can be calculated from the formula given below:

$$E = \sqrt{4kTBR}$$

where E = EMF in volts
    k = Boltzmann's constant which is 1.37×10^{-23} joules per degree kelvin
    T = absolute temperature in degrees kelvin
    B = bandwidth in hertz
    R = resistance in ohms

Unfortunately there is no way of reducing the noise generated by a resistor, except by reducing the resistance, the bandwidth or its temperature. It is for this reason that some very specialized very low noise amplifiers are cooled.

Signal to noise ratio

There are a variety of methods used to measure and define the sensitivity of a radio receiver. As the noise performance is the limiting factor, the sensitivity is specified in terms of its noise performance. The most obvious method of achieving this is to measure the difference between the wanted signal and the noise level under specified conditions as shown in Figure 5.54.

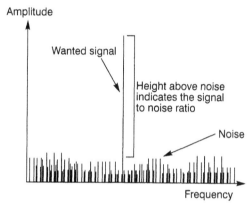

Figure 5.54 *Signal to noise ratio*

The difference between the wanted signal and the noise is expressed as a ratio in decibels. This is often termed the signal to noise or S/N ratio. The signal level also needs to be mentioned as this has a direct bearing on the figures obtained. The bandwidth of the receiver also has a bearing on the figures which are obtained because the noise level is proportional to the bandwidth being used.

When specifying the performance for AM the level of modulation also needs to be included in the specification. This is because the audio output from the receiver is measured to give the signal level and this will be dependent upon the modulation level. The usual value for modulation is 30 per cent.

A typical specification for a good short wave band communications receiver may be in the region of 0.5 microvolts for a 10 dB S/N in a 3 kHz bandwidth for SSB and Morse. For AM reception it may be in the region of 1.5 microvolts for a 10 dB S/N in a 6 kHz bandwidth with a 30 per cent modulation level. Note that a wider bandwidth is required for AM reception and this means that the noise level is higher and sensitivity is less.

In some instances a signal plus noise to noise (S + N)/N ratio is specified. The reason for this arises from the way the measurement is made. A signal generator is connected to the input of the receiver, and at the audio output a meter is used to measure the audio level. With the signal generator turned off the level of the noise is noted at the output. Then the signal generator output is turned on and its level is adjusted so that the output from the receiver is 10 dB higher than the noise level.

When the signal generator is turned on the audio level meter is reading the level of the output signal plus any background noise, i.e. (S + N)/N. Note that during the measurements for this test when the signal generator output is turned off the impedance seen by the radio receiver must be 50 ohms. This is normally true because the output attenuators and other matching circuits on the signal generator ensure the output is maintained at this value.

Occasionally the signal generator specification will mention that the signal voltage is given as a potential difference (PD) or the electromotive force (EMF). This is very important because there is a factor of two difference between them. The EMF is the open circuit voltage whereas the PD is the voltage under load as shown in Figure 5.55. If the load impedance is the same as the source impedance the EMF will be twice the PD at the load. In theory the EMF is a more correct way of specifying the voltage because the way in which the signal generator works assumes that it has been loaded with 50 ohms for the value to be correct. However, as the values with the PD look more impressive, it is this value which is normally stated. Even when EMF or PD is not specifically mentioned it should be assumed that PD is implied.

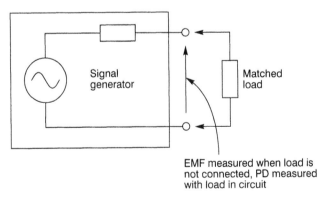

Figure 5.55 *EMF and PD in a signal generator*

SINAD

Sensitivity may also be specified in terms of SINAD measurements. This is very similar to a signal plus noise to noise measurement. The test method involves applying a signal modulated by a single tone. The audio from the receiver and the audio tone are notched out as required. In this way a measurement of signal plus noise plus distortion to noise plus distortion is obtained. The SINAD value is expressed in decibels.

Using this system the sensitivity of a set is normally quoted as a given number of microvolts to give a certain value of SINAD. A figure of 12 dB SINAD is normally used as this represents a 25 per cent distortion and noise factor.

SINAD measurements are most commonly used for FM sets. However, they can be used for other modes as well. For AM it is simply a matter of changing the type of modulation. For SSB it is necessary to ensure that the receiver is tuned to exactly the correct frequency so that the audio tone can be notched out. Once this has been done the measurement can be made in the normal way.

Noise factor and noise figure

Apart from signal to noise ratio and SINAD specifications, the noise performance can be specified in terms of a noise figure. This measurement is more versatile than the signal to noise ratio because it can be used to determine the performance of a piece of equipment whether it is a complete system, a receiver, or a smaller item such as a preamplifier. Essentially this measurement gives an indication of the level of noise which each item introduces.

The idea of noise figure and noise factor is built around the fact that in any system there is a certain amount of noise below which it is not possible to go and each element will introduce some noise over and above this. This thermal noise is the limiting factor and it is dependent upon the resistance of the system. A 50 ohm resistor generates a certain level of thermal noise. As an antenna looks like a resistor to the input of the receiver, this too generates noise of its own.

As a result of this thermal noise, any signal being picked up by the antenna has a certain signal to noise ratio associated with it. If the signal is very strong the signal to noise ratio will be good, but if it is weak it will be poor. When the signal is passed through an amplifier, receiver or any other piece of electronic equipment the signal to noise ratio will be degraded because the circuit will introduce additional noise.

The noise factor is determined by taking the signal to noise ratio at the input and dividing it by the signal to noise ratio at the output. For these calculations the signal to noise ratio must be given as a ratio and not expressed in decibels. As the circuit will always degrade the signal the noise factor is always greater than one. The noise figure is obtained by converting the noise factor into decibels as shown in Figure 5.56

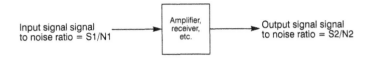

$$\text{Noise figure} = 10 \log_{10} \frac{S1/N1}{S2/N2}$$

Figure 5.56   *Noise figure of a system*

To give an example of this a signal may have a signal to noise ratio of 6:1 at the antenna, and this may be degraded to 4:1 after passing it through an amplifier. For this amplifier the noise factor would be 6/4 and the noise figure would be 10 $\log_{10}$ 1.5 or 1.76 dB.

Strong signal and overload

While the sensitivity of a receiver is very important it is not the only aspect which designers have to consider. Equally important is the way in which the radio is able to handle strong signals. A receiver may need to receive weak signals which are close to very strong ones. It is quite possible that a set may need to receive signals which vary in strength by up to 100 dB. This enormous variation in strength tests any design to its limits, particularly the front end stages.

Under normal conditions the amplifiers in a set must remain linear. In other words the output is directly proportional to the input. However, when strong signals are received there comes a point where the output of the amplifier starts to overload and cannot give out the required level. At this point the amplifier is said to be in compression.

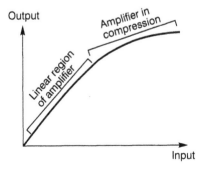

Figure 5.57   *Characteristic of a typical amplifier*

Compression itself is not a problem as the absolute values of a signal are rarely important and in any case the AGC acts on the signal level to alter the gain of the set, it is the effects associated with compression and non-linearity which cause the problems. Intermodulation, blocking, cross-modulation are three problems associated with overload, and these can noticeably reduce the performance of a receiver in some circumstances.

Intermodulation distortion

One of the main problems associated with overload is intermodulation distortion. When this occurs, new signals are generated within the receiver, sometimes giving the perception that it is receiving a lot of signals and it is very sensitive.

Two effects occur to give rise to intermodulation distortion when an amplifier becomes overloaded. One is that harmonics of signals being received are generated, and the second is that the amplifier, being non-linear, acts as a mixer. On their own these effects are unlikely to degrade the performance very much because the RF tuning would remove signals which would pass through the rest of the set. If a signal at frequency f is generating harmonics, these will fall at frequencies of $2f$, $3f$, $4f$, and so forth. For even the lowest harmonic to enter the set, the fundamental at half the frequency must enter the front end. The RF tuning will reduce this to a low level and it is unlikely to cause a problem for most applications. Similarly if two signals are to mix together to form a signal which is within the receiver pass-band they will be outside the acceptance range of the front end and their effects are not noticed under normal conditions. Being just a mix product between two signals this is called a second order effect.

The major problem occurs when harmonic generation and mixing occur together. It is possible for the harmonic of one signal to mix with the harmonic or fundamental of another to give a third signal which is within the pass-band of the receiver, i.e. $2f_1 - f_2$. It is only the difference products that cause a problem, as the sum products fall well away from the received frequencies. To give an example, signals on 100.00 and 100.1 MHz may enter the front end of a receiver. The harmonic of the first will be at 200.00 MHz and this can mix with the second (200 – 100.1) to give a signal at 99.99 MHz. All of these signals are well within the pass-band of the front end and the unwanted signal at 99.9 MHz will appear as a real signal. This is a third order effect. Other higher order products can be calculated and it can be seen that a comb like that shown in Figure 5.58 is produced. Note that only the odd order effects cause problems.

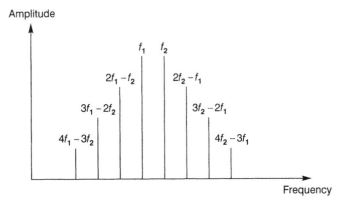

Figure 5.58 *Intermodulation products from two signals*

Third order intercept

To give an indication of the performance of an amplifier a figure known as the third order intercept point is often quoted. It is found that the levels of distortion in a circuit are very small under normal operating conditions, but rise very rapidly as the level of the input signal increases. For every 1 dB increase in signal level a third order signal will rise by 3 dB and a fifth order signal by 5 dB, and so forth.

Normally the amplifier would run into saturation well before the third order effect became comparable with the wanted signal. However, it is possible to plot the curves the levels of the two signals would take. At a certain point they intersect as shown in Figure 5.59. This is known as the

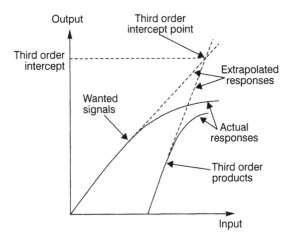

Figure 5.59 *Third order intercept point*

third order intercept point, and it gives an indication about its strong signal handling capacity. Typically a good professional high-performance receiver may have an intercept point of around 25 dBm (i.e. a signal which is 25 dB above a milliwatt).

Blocking

Often a very strong signal that is off channel can reduce the sensitivity of a receiver. This happens because the RF amplifier stage is being driven into compression. When this occurs it has the effect of only allowing the main signal through the amplifier, and all the others are reduced in strength. When blocking is quoted it is usually given as the level of signal 20 kHz away from the received channel which gives a 3 dB reduction in the wanted signal. Usually a good set will be able to withstand levels of several milliwatts before this happens.

Cross-modulation

Another effect caused by overloading is called cross-modulation. When this occurs the modulation or amplitude variations on a strong signal are superimposed on signals close by. It is particularly noticeable on AM signals where the modulation of a strong signal is superimposed onto weaker signals either side.

The problem is a third order effect, and often occurs as a result of poor mixer performance. However, any set with a high third order intercept point should give good cross-modulation performance as well.

Dynamic range

The dynamic range of the most important parameter of its specification as it outlines the range over which it can operate. In some instances it may be necessary for a set to be able to accommodate signals with enormous differences in strength, especially if a good antenna is used.

There are a number of ways in which the dynamic range may be determined. Essentially it is the difference between the weakest signal that it can hear and the strongest one it can tolerate without any noticeable degradation in performance. As these two end points can be specified in a number of different ways they must be included in the dynamic range specification.

The low end of the range is governed by the sensitivity of the set. Usually the minimum discernible signal (MDS) is used. This is the weakest signal that the set can hear, and it is usually taken as a signal equal to the noise produced by the set. Often the signal may be around −135 dBm for a 3 kHz bandwidth.

At the other end of the range there are two main limiting factors. One is the onset of blocking, and this may be at the point where sensitivity is reduced by either 1 dB or 3 dB. The other is the generation of intermodulation products. Here the point that is often taken is the level of input signals which generate signals which could mask a signal equal to the minimum discernible signal, i.e. the noise floor.

Typically sets have an intermodulation limited dynamic range of between 80 and 90 dB whereas the blocking limited dynamic range may be around 110 dB or more.

The dynamic range is usually limited by the performance of the front end stages of the set. Not only is the noise performance of these stages critical, but so too is the strong signal handling performance.

6 Transmitters

Today there is an enormous variety of different types of transmitter used to generate all the signals that can be found in the radio spectrum. Their purpose is to generate the basic signal or carrier, and then superimpose the modulation onto it in the correct format. Once this is done the signal is amplified to the correct level and filtered to remove any spurious products that are outside the required band. At this stage the output stages may also include matching circuitry to ensure that there is an accurate impedance match between the load and the transmitter. In this way the maximum power transfer takes place.

Today transmitters take a wide variety of forms. Usually they are paired with a receiver so that two-way communications can take place. Often units that contain a transmitter and receiver are called a transceiver. However, this term is normally reserved for commercial items where 'traditional' radio communications are required and items such as mobile phones that also contain both transmitter and receiver are not referred to by this term. Whatever the use of the transmitter, the fundamental concepts are the same, and the same building blocks are used and the same specifications apply.

Transmitter building blocks

Like all electronics the complexity of transmitters is increasing, and their flexibility is improving to enable them to fulfil the increasing requirements placed on them. From low power UHF transmitters in cellular phones to the high power broadcast stations, they all use a number of the same basic building blocks to achieve the output of the required signal.

Oscillator

The heart of any transmitter is its master oscillator. This generates the carrier onto which the modulation is superimposed. Many transmitters utilize a number of oscillators to mix the signal to its final required frequency. However, this processing often takes place after the modulation has been applied.

The requirements of the oscillator wherever it is used in the equipment are that it should be stable and not drift. If it does then the output signal will change. In some instances drift on an oscillator may cause the characteristics of the signal to change as in the case when single sideband is used.

The oscillators used in transmitters take many forms. Those transmitters that do not need a frequency change may use a crystal oscillator. This will ensure a high degree of frequency stability. Typically this approach is adopted where small low power transmitters are used. In some cases where high degrees of accuracy and stability are required, crystal ovens may be employed.

For signals which need to change in frequency a variable frequency oscillator (VFO) is required. The simplest form of VFO uses an inductor and variable capacitor. An example of a suitable oscillator is shown in Chapter 5, Figure 5.31. The output from this circuit must be buffered using a further amplifier. This will reduce the effects of any changes in load that may occur and cause frequency shifts. Buffer amplifiers generally have a high input impedance to ensure that the oscillator circuit is not loaded to any degree.

In some instances more sophisticated oscillator systems may use a crystal mixer system. Here a VFO running at a relatively low frequency is

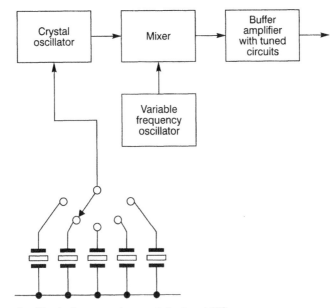

Figure 6.1 *Block diagram of a crystal mixer VFO*

mixed with a crystal oscillator running at a higher frequency. If wideband operation is required the crystals in the crystal oscillator can be switched to increase the frequency coverage. The inherent stability of the crystal oscillator enables the whole oscillator system to remain stable despite the fact that part of the circuit is switched. Also by keeping the variable oscillator running at a relatively low frequency the stability is maintained. However, the output of a VFO which uses a mixer must be tuned to prevent unwanted or spurious signals from entering the later stages of the transmitter, and ultimately being radiated. This approach is seldom used these days.

Today most transmitters that need variable frequency capability use a frequency synthesizer. This approach gives much greater stability while still being able to operate over a wide range of frequencies. Even where a number of preset frequencies may be required, the technology is so cheap these days that synthesizers offer a very effective way of generating the required signal in a transmitter.

Synthesizers for use in transmitters are exactly the same as those employed in receivers. In transceivers the same one is used for both transmit and receive functions. Synthesizer phase noise is also important in transmitters. In the same way that receivers suffered from reciprocal mixing as a result of phase noise generated by a synthesizer, a poorly designed synthesizer will generate wideband noise extending either side of the signal. This will be transmitted and may cause interference to other users nearby if the signal is strong enough. For data transmissions using forms of phase modulation, high levels of phase noise may introduce data errors, increasing the bit error rate. As a result care must be taken in the design to ensure that this does not occur.

Mixers

The basic mixing process and the circuits used have been described in Chapter 5. Not only are mixers used widely for receiver applications, but they are also widespread in transmitter circuitry as well. Here they are also used in frequency conversion applications to change a signal from one frequency to another. In addition to this they are used in the modulation process. For example, they can be used in the generation of the basic signal being used when phase modulation or single sideband is required. When mixers are used in frequency changers, care must be taken to ensure that the unwanted mix products are removed so that these are not transmitted otherwise spurious signals will be transmitted and these are likely to cause interference to other users of the radio spectrum.

Amplifiers

Once the signal leaves the area of the transmitter where it is generated it is normally necessary to amplify it to increase the power to the required level to be applied to the antenna and be transmitted. After the signal has been generated its power level may only be a few milliwatts, whereas the required level is likely to be much higher than that. For example, high power broadcast stations may have output powers of tens or even hundreds of kilowatts. Many other transmitters run much lower powers. Automobile-based systems like those used for commercial point to point communications may have output powers of around 10 watts, whereas mobile phones may develop a power of a watt or even much less. To achieve these power levels amplifiers are required to bring the output power to the required levels.

Small transmitters including hand-held transceivers, mobile phones and the like use amplifiers that are semiconductor based. Even many high power transmitters still use semiconductors, but those which deliver many kilowatts use thermionic technology (valves or tubes) as they are still the most convenient way of generating these high levels of RF power.

An RF amplifier is like any other form of amplifier having the same basic requirements and constraints. The only difference is that the frequencies at which the amplifier operates may be higher. Essentially a signal enters at the input and an amplified or larger version appears at the output. Where the signal being amplified contains elements that are dependent upon the level of the signal, i.e. amplitude modulated components, then the amplifier must be as linear as possible. Here the output level must be a linear function of the input, i.e. multiplied by a fixed number regardless of the input level. For systems running CDMA such as mobile phone systems linearity is again a requirement.

In some transmitter applications this may not be the case. Good linearity is accomplished at the expense of poor power efficiency. In some cases efficiency is very important. For portable applications it is necessary to ensure that as much of the DC power entering the amplifier is converted into radio frequency power that can be transmitted. In this way the batteries can be made to last as long as possible. At the other end of the spectrum broadcast stations with transmitters that consume many hundreds of kilowatts find that the cost of the electricity is a major consideration. By increasing the efficiency of the amplifiers, significant cost savings can be made.

The efficiency of the amplifier is simply the proportion of the DC input power supplied to the amplifier that is converted into radio frequency power to be transmitted. This can be expressed:

$$\eta = \frac{\text{Power output}}{\text{Power input}} \times 100\%$$

There are a number of factors that affect the efficiency of an amplifier. One of these is the mode of its operation. Those that operate in a totally linear mode are less efficient than those that run in a less linear mode. As a result amplifiers are classified in terms of their mode of operation. Normal linear amplifiers are said to operate in class A. They are biased to run in the linear region of the amplifying device whether it is a bipolar transistor, FET or even a valve (tube). They are less efficient than other types of amplifier because they conduct current for the whole of the input signal cycle as shown in Figure 6.2. The maximum possible efficiency for a class A amplifier is 50 per cent, with values between 25 and 45 per cent being more normal.

Efficiency can be improved by biasing the amplifier to conduct for only part of the cycle. For what is termed class B operation the amplifier conducts over only half the cycle as shown in Figure 6.3. In this case the amplifier is biased at the cutoff point of the device. If an amplifier of this type is operated into a resistive load it effectively rectifies the signal,

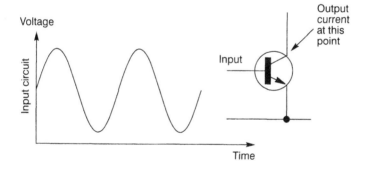

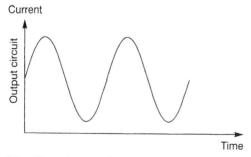

Figure 6.2 *Class A operation*

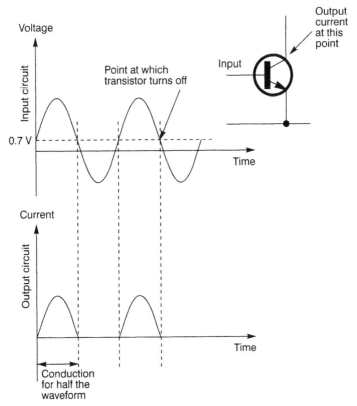

Figure 6.3 *Class B operation*

passing only one half of the cycle. Sometimes this type of circuit even may be used as a simple detector. Another application is to use two class B amplifiers operating over opposite halves of the cycle. By summing the outputs both halves of the cycle are present, and a linear representation of the signal is produced. This type of operation is called push–pull, because each half of the amplifier operates on a different half of the cycle.

For RF applications a single ended class B amplifier may be used in conjunction with a tuned circuit. In operation the tuned circuit 'rings' when the half wave signal is applied, and this supplies the missing half of the cycle. Amplifiers of this type typically have an efficiency of around 60 per cent.

To improve the efficiency still further, the amplifier may be operated in class C. In this mode of operation the amplifier is biased beyond the cutoff point of the device. In this way it only allows short current pulses of less than half a cycle to flow, corresponding to the peaks in the input

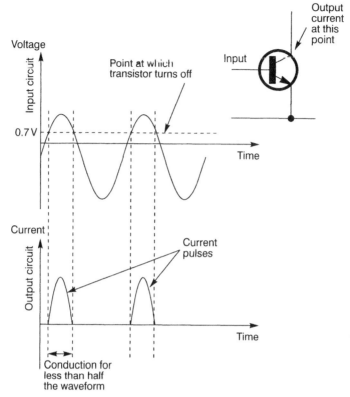

Figure 6.4 *Class C operation*

waveform. In view of their mode of operation class C amplifiers are very non-linear and the output is rich in harmonics. This means that a filter is required to remove them before being passed to an antenna. The filter is tuned to the frequency of operation and only allows through the required frequency. In this way the required sine wave is extracted. In view of the fact that the amplifying device is biased beyond its cutoff point higher levels of drive are required for this type of amplifier when compared to a class A or class B amplifier.

Class C amplifiers perform very well as frequency multipliers. Their output is particularly rich in odd harmonics, extending to many times the fundamental frequency. However, it is normal to select only the third or at the highest the fifth harmonic. This is achieved using the filter in the amplifier output circuit. If higher order harmonics are selected it is not normally possible to achieve sufficient rejection of the unwanted harmonics, and this results in unwanted signals being transmitted.

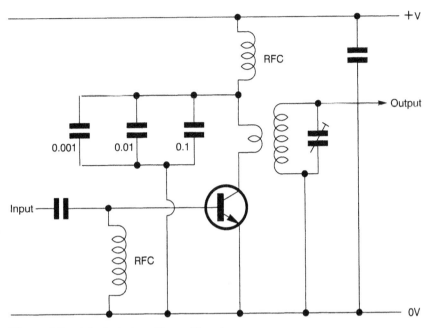

Figure 6.5   *A typical class C amplifier stage*

One very important aspect of a transmitter power amplifier is the amount of power that it can deliver to the antenna system. Radio frequency power is measured in exactly the same way as low frequency AC power. It is simply the voltage times the current, assuming they are in phase with one another. Assuming that the antenna load is resistive then they will be in phase. Very high power amplifiers like those used as the final stage for a broadcast transmitter may be capable of delivering 100 kilowatts or more to the antenna, whereas those used for local broadcast stations may only deliver 100 watts or so. Other types of transmitter can deliver a variety of power levels dependent upon their use. Cellular phones can typically deliver around a watt, but this will vary according to their application and the type of system in use.

Power levels are often measured directly in watts. However, they are often specified in decibels relative to a watt or a millliwatt. Given the units dBm (decibels relative to a milliwatt) and dBW (decibels relative to a watt), they are being used increasingly. Using this system it is very easy to incorporate any changes in level brought about by the use of amplifiers, attenuators, losses in feeders, etc. All of these are generally expressed in decibels so it is simply a matter of adding or subtracting the relevant figure from the power level, e.g. a power of 10 dBW becomes only 6 dBW at the end of a feeder with a 4 dB loss. For transmitters, the

Table 6.1 *dBm–dBW–watts conversion*

dBm	dBW	Watts	Terminology
+100	+70	10 000 000	10 megawatts
+90	+60	1 000 000	1 megawatt
+80	+50	100 000	100 kilowatts
+70	+40	10 000	10 kilowatts
+60	+30	1 000	1 kilowatt
+50	+20	100	100 watts
+40	+10	10	10 watts
+30	0	1	1 watt
+20	−10	0.1	100 milliwatts
+10	−20	0.01	10 milliwatts
0	−30	0.001	1 milliwatt
−10	−40	0.0001	100 microwatts
−20	−50	0.00001	10 microwatts
−30	−60	0.000001	1 microwatt
−40	−70	0.0000001	100 nanowatts
−50	−80	0.00000001	10 nanowatts
−60	−90	0.000000001	1 nanowatt

power levels are generally expressed in dBW, whereas levels in dBm are generally used for low power circuits. For example, a diode ring mixer may require a local oscillator input of 7 dBm.

It is important to ensure that an amplifier operates under the correct conditions. When there is a poor impedance match between the amplifier and its load it will operate under conditions that are not ideal. Amplifiers are designed to operate into a given load. This is normally 50 ohms for which suitable feeders are available. While the feeder itself may be perfectly matched to the amplifier, it is possible that the antenna or other load connected to the remote end of the feeder may present a poor match. When this happens power is reflected back to the amplifier and high levels of VSWR may be present (see Chapter 4). Under these conditions damage may occur to the amplifier because high levels of current or voltage may be present. If a high voltage occurs, this may cause the maximum voltage for the device to be exceeded with resultant break-down of the device. Alternatively high current levels can cause its current ratings to be exceeded resulting in the burnout of the device. Fortunately it is possible to design low power amplifiers to withstand poor load conditions, but this is not usually possible for high power amplifiers. In view of this many high amplifiers incorporate protection circuitry which reduces the power level as the SWR increases. Other protection systems may remove the power from the amplifier if a high level of SWR is seen. If protection circuitry is required, then it must be very fast acting. Output transistors can be destroyed very quickly, especially if voltage breakdown occurs.

Filters and matching networks

The networks used to filter and match transmitters are very important to provide the correct matching from an amplifier to the next stage or into the antenna. By providing the correct impedance match, the greatest power transfer is effected. As the matching networks consist of inductors and capacitors, they are usually tuned to ensure that any unwanted signals are reduced to keep them within the required limits.

A variety of different coupling methods or networks can be employed. The actual choice is dependent upon a number of considerations including the available driving power, the selectivity required, the impedances being matched and the level of mismatch that can be tolerated. Often the input impedance to the base of a transistor amplifier may be as little as 10 ohms falling to an ohm or less as powers rise above a watt or so. Accordingly not all types of LC matching networks are suitable for these circumstances. Also in some circumstances a mismatch is introduced into the design to control the power distribution in the amplifier chain and to aid stability. The Q of the tuned circuits is another important factor. This is the 'quality' factor of the circuit and is determined by taking the bandwidth and dividing it by the resonant frequency of the circuit. While a high Q is desirable to reduce the level of spurious signals, many transistor designs use low Q circuits to reduce the possibility of instability. Most solid stage amplifiers use matching networks with loaded Q levels of five or less. Those amplifiers that use valve or tube technology, and a few are still in use, can use higher levels of Q, typically between 10 and 15 because of the higher grid circuit impedances. To indicate the type of circuits that may be used, the circuits of typical amplifiers using discrete components are shown in Figure 6.6. In this, first a transformer is used to give the required impedance transformation whereas in the second a capacitive divider is used. It can be seen that there is a high level of decoupling provided on the supply to prevent high levels of RF energy appearing on the common supply used by all circuits in the transmitter.

Many of the spurious signals emanating from an RF amplifier are harmonics of the wanted signal, especially when the amplifier is run in class C. To remove them the final amplifier often uses a low-pass filter after any matching circuitry. The general format for these filters is shown in Figure 6.7. It can be seen that the filter consists of series inductors with

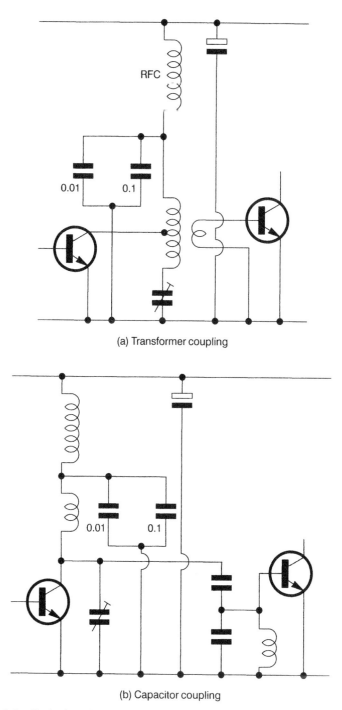

(a) Transformer coupling

(b) Capacitor coupling

Figure 6.6 *Typical methods used to couple RF amplifiers in a transmitter*

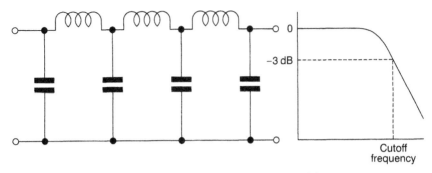

Figure 6.7   *An example of a low-pass filter with a typical frequency response curve*

capacitors running to ground. The filter has a cutoff point where the response has fallen by 3 dB. This is chosen to be above the frequency of operation.

The response of the filter can be controlled to a degree by changing the relative values of the components. Factors such as in-band ripple, the initial roll-off, phase characteristic and so forth can all be varied, and names are given to the different types of filter. A Butterworth filter gives the flattest in-band response, for example. The values are not simple to derive manually, but they can be obtained either from books that contain tables of values that can then be scaled for the required frequency and impedance, or by using a computer program.

These filters are often described as having a certain number of poles. In general terms a filter has a pole for each capacitor or inductor it contains. In other words the filter shown in Figure 6.7 has seven poles. The greater the number of poles the faster the filter response rolls off beyond its cutoff point.

Unfortunately it is not possible to produce the perfect filter and this means that small levels of unwanted signals will always be present. However, they must be reduced to a level where they will not cause any undue interference to other users. When analysing a signal for harmonics or other spurious signals a spectrum analyser is used. This gives a plot of amplitude normally on a logarithmic (decibel) scale against frequency, enabling signals to be seen and analysed. A diagram of a typical plot from a spectrum analyser might look like that shown in Figure 6.8.

In this example it can be seen that the main signal falls at a level of −10 dB, and the next strongest signal is at −40 dB. This means that the second signal is 30 dB below the main one.

When stating the level of a spurious signal the levels are not normally given in absolute levels, i.e. a given number of watts or milliwatts. The more usual way is to relate them to the level of the wanted signal. In other

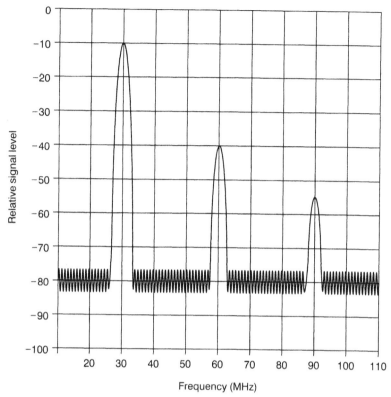

Figure 6.8 *Diagram of a typical spectrum analyser screen plot*

words a spurious signal will be said to be a certain number of decibels below the carrier. Sometimes this will be referred to as a certain number dBc. In the previous example the figure could have been given as −30 dBc. The more negative the number is the smaller the spurious signal level. In other words a spurious signal level of −40 dBc is better or lower in level than one of −30 dBc. This method is convenient to use because it is easy to see the difference between the main signal and the spurious signal on a spectrum analyser.

Intermodulation

In just the same way that non-linearity of a receiver amplifier can cause intermodulation distortion, the same can be true of an amplifier in a transmitter. In many cases non-linearity itself is not a problem, for example where the signal is frequency modulated, but where the signal is

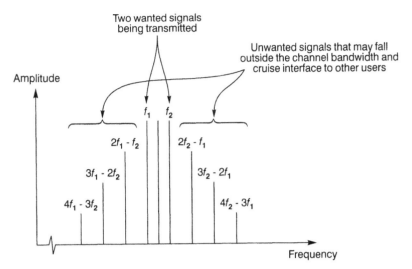

Figure 6.9   *Intermodulation products produced in an amplifier*

amplitude modulated, either as AM or SSB, etc., then it will give rise to unwanted products. The problem is encountered where a linear amplifier is used on the output of an AM or SSB transmitter or when other types of modulation such as quadrature amplitude modulation (QAM) are used.

An AM or SSB signal consists of a variety of different signals dependent upon the audio being applied. However, to illustrate the problem the example of two signals being transmitted will be taken as an example. This might happen when using an SSB transmitter where just two audio tones, one at 1 kHz and another at 2 kHz, are being transmitted. Here the third order products will produce signals at $2f_1 + f_2$ and $2f_2 + f_1$. In the case of the example the two third order intermodulation products will appear 1 kHz either side of the two main signals, higher order mix products, i.e. $3f_1 + 2f_2$ and $3f_2 + 2f_1$, will appear a further 1 kHz away, and so forth as shown in Figure 6.9.

In the case of a signal modulated with speech or music, there will be a whole variety of different audio frequencies making up the waveform. All these various frequencies intermodulate with one another to generate noise or splatter which spreads out from the main signal. Normally the worst intermodulation products will be those which are nearest to the wanted signal, and their levels reduce as the offset increases.

The specifications for intermodulation products are usually given in terms of the difference between the wanted or main signal and the various intermodulation products. This figure is expressed in decibels. Often a transmitter specification will say that all intermodulation products are below a given level. In this case the worst ones are bound to be the third order products. Sometimes the levels of specific products will be stated. Typically the third order products will be around −25 to −30 dB for the third order products and 5 or 6 decibels lower in the case of the fifth order products.

Spurious outputs

Intermodulation products are one type of spurious signal. Other types can be generated in transmitters typically as unwanted mixer products or as harmonics. These may appear on frequencies that are far away from the wanted signal. It is very important that a transmitter does not radiate these unwanted signals otherwise they could cause serious interference to other users of the frequency spectrum. Accordingly it is necessary to ensure that they are kept to acceptable levels and this is done by ensuring that filters are used within the designs themselves as well as at the output. While it is never possible to completely eliminate spurious signals their levels are kept to limits where they are deemed not to cause a problem and their levels are often specified in the data sheet for the equipment. When specifying the levels they are given as levels in decibels relative to the main signal or carrier. For example, the harmonics may be better than −50 dB and other spurious signals better than −55 dB.

Output impedance

The output impedance of a transmitter is a particularly important parameter. Some transmitters will be connected to a feeder, possibly coaxial to transfer the power to an antenna at some distance from the transmitter. Others such as those used in cellular telephones feed directly into the antenna that is part of the overall telephone assembly. Whatever the system it is necessary to ensure that good match is achieved between the output of the amplifier and the load, to enable the output amplifier to operate correctly and enable the optimum power transfer to take place.

The output impedance of the transmitter is measured in ohms. As might be expected it relates to the ratio of the voltage and current provided by the output. For those transmitters that are connected to a feeder, the output impedance is almost invariably 50 ohms as this is a standard impedance for coaxial feeder. Placing an incorrect load on the transmitter can result in damage. High voltages or currents can destroy the output device. Although very low output units may be able to operate into any load, those above a few watts can be damaged, and often protection circuitry is provided. Running the transmitter with a high level of standing wave ratio in a feeder has exactly the same effect. Depending upon the distance from the load, the transmitter may see an excessive level of current or voltage. In view of this the output impedance of a transmitter is often quoted, and the maximum level of standing wave ratio may also be given.

Simple Morse transmitter

To give an idea of how various transmitters operate, overviews of a number of types of transmitter are given, starting with a simple example to give an idea of the basic requirements of a transmitter. An effective Morse transmitter can be made out of very few components. As Morse signals consist of an oscillation that is keyed on and off, the circuitry can be kept very simple. Indeed it is possible to make a Morse transmitter with only two or three transistors and a handful of other components. However, more sophisticated designs can naturally be made if required.

Figure 6.10 shows the block diagram of the very basic Morse transmitter. All that is required is an oscillator to generate the signal, a power amplifier to increase its level, a method of switching the signal on and off, and finally the matching and filtering circuitry at the output.

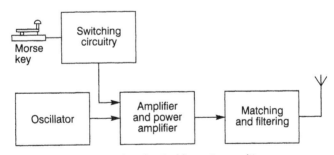

Figure 6.10   *Block diagram of a simple Morse transmitter*

In the very simplest circuits the oscillator may be crystal controlled to ensure sufficient stability, although this does limit the flexibility of the set. Often a VFO is employed to enable the set to operate over a band of frequencies. This could be an LC tuned oscillator or a more complicated unit based around a synthesizer. However, for a transmitter of this simplicity it is unlikely that a sophisticated synthesized oscillator would be used.

The power amplifier is used to increase the level of the signal. In this type of transmitter where no analogue modulation is applied to the carrier, the PA will be operating in class C to give the maximum efficiency. The final stage is the filtering and matching circuitry. For a low power transmitter it is often possible for the output device to give a good match to 50 ohms with little or no impedance transformation. As a result only filtering may be required. For a transmitter of this nature this will consist of a low-pass filter. This will remove the spurious signals that will be harmonics of the required signal. Typically this may be a five pole filter with a cut-off just above the operational frequency.

Amplitude modulation transmitter

A huge number of signals in the radio spectrum are modulated to carry audio. The simplest way of achieving this is to modulate the amplitude of the carrier in line with the audio signal. While this method is not as efficient as many other methods, it is still widely used for broadcast stations in the long, medium and short wave bands as well as being used by aircraft at VHF.

A basic AM transmitter requires the carrier to be generated and this is modified by the modulating signal. Figure 6.11 shows a basic AM transmitter. In this circuit a system called high level modulation is used, in other words the modulation is applied to the carrier in the final amplifier.

The carrier is generated by the oscillator circuit. This may be a variable frequency oscillator, or possibly a crystal oscillator. Today frequency synthesizers are often likely to be used, for convenience and flexibility.

Once the basic signal has been generated it is buffered and amplified. This brings it to the correct level for the signal to drive the final power amplifier. Both the drive amplifiers and the final amplifier will operate in class C to maintain the optimum efficiency.

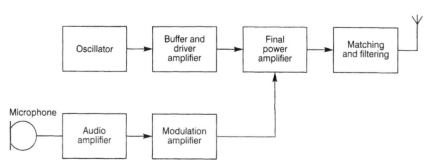

Figure 6.11 *A basic AM transmitter*

The audio signal from the microphone or other source enters the transmitter and is amplified. These stages may also limit or process the signals to improve their intelligibility, or reduce the bandwidth occupied. For most communications purposes frequencies above 3 kHz and below 300 Hz are attenuated. For broadcast purposes top limits above this are used dependent upon the channel spacing in use. On the short wave bands this is 5 kHz and on the medium wave band it is 9 kHz in Europe and 10 kHz in North America.

The audio signal is further amplified by a driver amplifier if required and applied to the modulation amplifier. This is a high power audio amplifier that must be capable of developing an audio power equal to half the input power requirements for the final RF power amplifier if 100 per cent modulation is to be achieved. In other words if the RF amplifier consumes 100 watts, then the audio amplifier must be able to deliver 50 watts. The power audio signal is then used to modulate the supply to the final RF power amplifier and in this way modulate the RF signal.

If any further increase in the power level of the transmitter is required after the modulation has been applied, the RF amplifier must run in a linear mode, typically class A or a variant of class A to prevent distortion of the modulation. These amplifiers are often called linear amplifiers in view of their mode of operation.

The final stage of the transmitter, like that of the Morse transmitter, is to pass the signal through the matching and filtering circuits. Again these are required to reduce the levels of spurious signals to acceptable limits.

Single sideband transmitter

There are a number of different methods in which a single sideband signal can be generated. The first and most commonly used employs a filter to remove the unwanted sideband. A second method uses phasing techniques to eliminate the unwanted elements in the signal.

The filter method is virtually a reverse of the signal path followed in a typical receiver. This is the most common method used as it is capable of giving excellent results. It is also very convenient for use in transmitter–receivers (transceivers) because many of the circuits can be used in both the transmit and receive paths, saving on production costs.

The outline of an SSB transmitter using the filter method is given in Figure 6.12. The first block in the transmitter is the carrier oscillator. This is normally crystal controlled because its frequency stability is critical to maintain the correct frequency relative to the filters used later.

Audio signals from the microphone are amplified and processed as necessary and applied with the carrier to a balanced mixer circuit. The

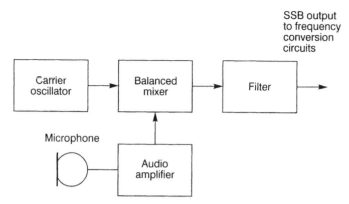

Figure 6.12 *Block diagram of an SSB generator using the filter method*

advantage of using this type of mixer is that the input signals are suppressed at the output. This generates a double sideband suppressed carrier signal. In other words the carrier, being one of the input signals, is suppressed by this circuit.

Once this signal has been generated, it is passed into a filter to remove the unwanted sideband. The filter will typically have a level of selectivity sufficient to pass the required audio bandwidth. As single sideband is usually used for communications purposes the filter will be relatively narrow, typically the bandwidth will be about 2.7 kHz or slightly more.

The position of the carrier frequency relative to the filter pass-band is important. It is necessary to ensure that the wanted sideband is not attenuated while ensuring that no undue amounts of the unwanted sideband are passed through. Normally the filter and carrier are positioned as shown in Figure 6.13.

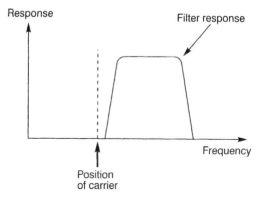

Figure 6.13 *Frequency of carrier and filter relative to one another*

It is normally found that the filter will give additional attenuation to the carrier, enabling high degrees of attenuation than the mixer alone can provide. For most transmitters the carrier will be suppressed by at least 40 dB as will the unwanted sideband.

The single sideband signal is generated on a fixed frequency. Most transmitters will need to be used on a variety of frequencies. To enable this to be achieved the superhet principle is used in the reverse way to that employed in a receiver. The fixed frequency single sideband signal is mixed with a variable frequency local oscillator to give a mix product on the required frequency. As in the case of a receiver there are a number of methods of achieving this. A single variable frequency oscillator on its own is unlikely to give sufficient stability and therefore virtually all sets now use a frequency synthesizer to give the stability and coverage without the need for band switching.

When mixers are used it is very important to remove the unwanted mix products. If not they will pass through the later stages of the transmitter and it will be difficult to remove them all in the final stages of filtering. As such it is good practice to place good filters after each mixer to ensure that the unwanted signals are removed as soon as possible after they are generated.

Apart from converting the signal to its final frequency the signal also needs to be amplified to the required level. In view of the fact that the signal carries analogue modulation the amplifier must preserve the nature of the modulation and must be linear. Any distortion will result in the audio becoming distorted as well as the signal occupying a wider bandwidth and causing interference to others.

The second method of generating a single sideband signal is to use phasing techniques. Although not widely used today it had a number of advantages as a cheap and easy method of producing an SSB signal in some applications in the past. A block diagram for this method of SSB generation is shown in Figure 6.14. The audio and carrier are split into components shifted by 90 degrees and applied to balanced modulators. The outputs from these are combined, at which point one sideband is reinforced while the other is cancelled out. The reverse sideband can be selected by transposing the audio or carrier phasing.

Using this method of SSB generation it is possible to generate an SSB signal at the operating frequency. However, it is necessary to maintain very accurate control of the amplitude and phase of the signals if cancellation of the reverse sideband is to be achieved. Even a 1 degree change in the phase of either the RF or audio signals will reduce the sideband suppression to 40 dB if a perfect match of the amplitude of both signals is maintained. Similarly mismatches in the levels will cause the suppression of the unwanted signals to be reduced. This limits the bandwidth of operation quite considerably. The generation of a 90 degree

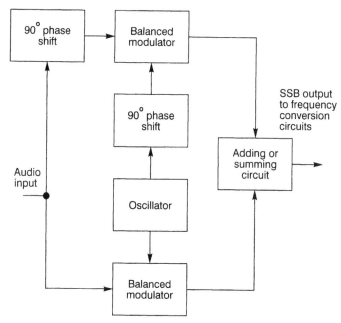

Figure 6.14   *Phasing method of SSB generation*

phase shift over the complete audio bandwidth requires some innovative circuit design. Often high tolerance components are required to achieve this. Another solution is to use a Gingell polyphase shift network. Using this it is possible to obtain the correct phase shift using standard tolerance components. In view of the ease with which the filter method can be implemented, and the flexibility of frequency synthesizers to give tuning, the phasing method of SSB generation has few advantages to offer. Accordingly it is not widely used today except in some limited applications.

Frequency modulation transmitter

Frequency modulation is used for a variety of applications from wideband FM broadcasting to handheld transceivers using narrow-band FM. There are a variety of circuits for producing an FM signal, and these depend on the type of modulation, whether it is narrow bond or wideband and the circuits being used.

A simple FM transmitter may take the form shown in Figure 6.16. Here the modulating signal is applied directly to the oscillator. This can be accomplished as shown in Figure 6.17. The audio signal applied to the

Figure 6.15   *A hand-held FM transceiver*

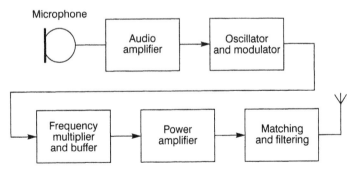

Figure 6.16   *A basic FM transmitter*

circuit changes the capacitance of the varactor diode. As the crystal is operating in its parallel resonant mode its frequency can be changed by external components. In this way the frequency of oscillation can be changed, if even by a relatively small amount.

FM transmitters are usually used for frequencies above 30 MHz and as a result crystals rather than LC tuned variable frequency oscillators are

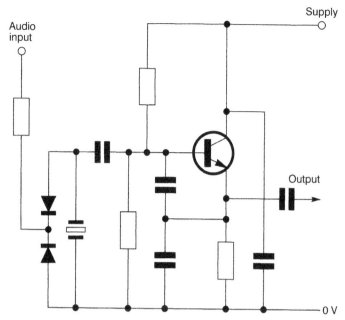

Figure 6.17 *Applying frequency modulation to an oscillator*

normally needed to provide sufficient stability for these basic transmitters. The usually way in which they work is for the oscillator to run at
a low frequency, typically a few megahertz, and this signal is multiplied
in frequency by a series of multipliers. These are normally amplifiers
running in class C to produce high levels of harmonics. Tuned circuits in
the output circuit select the correct harmonic. Normally these multipliers
are restricted to relatively low multiplication factors, two and three being
the most common.

It is found that as the frequency is multiplied so the level of deviation
is also increased. This means that if a signal is multiplied 18 times before
being transmitted, the level of deviation applied to the oscillator only
needs to be relatively small. For a final deviation of 3 kHz, the deviation
needs to be 3/18 = 167 Hz for a multiplication factor of 18. This can be
achieved relatively easily using the circuit shown.

Applying audio directly to the oscillator is not the optimum method of
generating FM. The level of FM can vary from one crystal to the next if the
specifications of all the crystals are not exactly the same. Also the linearity
of the conversion is not always good. This can be improved by applying
a fixed bias to the varactor diode to overcome its non-linearities.
However, a superior method is to use a phase modulator placed after the
oscillator as shown in Figure 6.18. The actual frequency deviation given

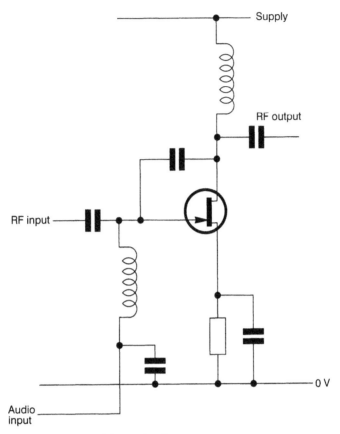

Figure 6.18   *A phase modulator circuit*

by the phase modulator increases with the increasing audio frequency at a rate of 6 dB per octave. The phase modulator can be made to give an FM compatible signal altering the audio response of the audio amplifiers to give the inverse response. The use of a phase modulator of this nature is far more satisfactory than modulating the oscillator. Sometimes this method of modulation is known as indirect FM as shifting the phase of a signal gives a corresponding change in frequency.

Once the FM has been generated it is amplified (and frequency multiplied) to the correct level (and frequency). Class C amplifiers can be used because all the information for the modulation is contained within the frequency changes. There should not be any amplitude variations. The filtering stages are particularly important for this type of transmitter where a large number of harmonics are generated. Each stage should contain sufficient filtering to ensure that the levels of spurious signals are

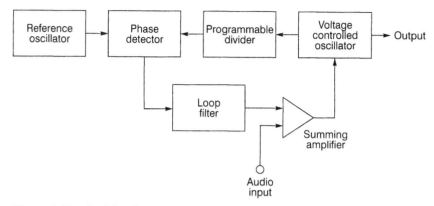

Figure 6.19 *Applying frequency modulation to a frequency synthesizer*

kept to an acceptable level. Then at the output the filtering and matching circuit must ensure a good match to the antenna system as well as keeping any unwanted signals from being radiated.

Instead of applying the modulation to a comparatively low frequency signal that has its frequency multiplied, a frequency synthesizer may be used to generate the signal. Modulation may then be applied to the control input of the VCO. Provided the loop bandwidth is less than the lowest audio frequency required then the modulation will appear on the signal. When a synthesizer is used the signal is normally not multiplied. Instead sufficient modulation can be applied to provide the required level of modulation.

7 Broadcasting

Most people first come into contact with radio by listening to radio broadcasts. Today many households have a variety of broadcast receivers from portable sets for use around the home, and car radios to very expensive hi-fi systems. These are used to pick up broadcasts on the long, medium and short wave bands as well as the higher quality transmissions on the VHF FM band. In addition to this new digital broadcast systems are being introduced and these are capable of providing even higher quality reproduction than that which can be obtained with VHF FM. Apart from providing improved quality these transmissions can also provide a new range of facilities as they are able to transmit data alongside the audio.

AM broadcasts

The first broadcasts to be made used amplitude modulation, and although better methods of modulation are available today, AM is still in widespread use on the long, medium and short wave bands. There are a number of reasons for this. The first is partly historical. Having become well established, with a large number of transmitter networks and many millions of receivers it is not possible to discontinue the service easily. Receivers for AM can also be made very cheaply and easily. Another factor is that the frequency allocations have many advantages in terms of coverage and propagation. However, these frequencies are not suitable for FM in view of the propagation characteristics as the multipath effects resulting from ionospheric propagation would lead to severe distortion. There is also insufficient spectrum available on these frequencies to allow for the high quality wideband FM transmissions.

The long wave band is not available for broadcasting in all areas of the world, although the medium wave band is a worldwide allocation. In view of the quality which can be attained this modifies the nature of the broadcasting. In areas where the stations are competing with FM services the medium wave tends to have a high percentage of speech-based programming. However, this does not mean that music programming is not present, although stations using the medium wave band are finding that listeners are migrating to the high quality FM broadcasts.

A variety of transmitter powers are used. In the UK, for example, the BBC operates a number of national networks. There are also independent national networks as well. Transmitters for these run many kilowatts and can be heard over large distances. In addition to these there are many local stations. These are only intended to have a relatively small coverage area and the transmitters may only run 100 watts or so. In North America it is normal for transmitters to be reduced in power at night when propagation via the ionosphere is possible.

In recent years the broadcast bands have been standardized so that stations transmit on particular channels. The channel spacing on the medium wave band is 9 kHz in Europe and 10 kHz in the USA. This reduces the amount of interference because the number of annoying heterodynes from stations 1 or 2 kilohertz away from one another are eliminated.

Although many new forms of broadcasting are available these days short wave broadcasting is still very popular. It is cheap to receive the transmissions as short wave radios are widely available for reasonable prices and many people, especially those in remote areas, still use this means for reception. In countries in Europe and North America other forms of broadcasting are more widely used and short wave reception is mainly by enthusiasts.

On the short wave bands stations are almost exclusively used for international broadcasting. Because of the nature of propagation at these frequencies, it is unlikely that these stations will achieve much coverage within their own country, unless the skip distance falls within the country. As a result short wave stations are used as flagships for their countries. Often they carry propaganda. This was particularly true in the days of the cold war. Stations from the West could be heard giving exactly the opposite views to those from the Eastern Bloc. Even today short wave broadcast stations are still used for this purpose. However, many stations seek to give their countries credibility on the international scene by producing good reliable programming and news. The BBC World Service is recognized the world over as possibly the best station. There are also many religious stations on the air including stations like Vatican Radio, and HCJB which broadcasts from the Andes.

Sometimes the short wave bands are used for domestic broadcasting. Many of the countries in the tropical areas of the world are large and relatively sparsely populated. This means that it is not possible to obtain adequate coverage with either medium wave or VHF FM transmissions. In countries like the UK 20 or more stations may be required to give reasonable coverage for a national network. In countries which are possibly larger and are less densely populated it is not economically

Table 7.1   *Broadcast allocations up to 30 MHz*

Long wave	0.150–0.285
Medium wave	0.5265–1.6065
120 metres	2.300–2.495*
90 metres	3.200–3.400*
75 metres	3.900–4.000**
60 metres	4.750–5.060*
49 metres	5.950–6.200
41 metres	7.100–7.300**
31 metres	9.500–9.990
25 metres	11.650–12.050
22 metres	13.600–13.800
19 metres	15.100–15.600
16 metres	17.550–17.900
13 metres	21.450–21.850
11 metres	25.670–26.100

All frequencies are in MHz.
* Tropical bands only for use in tropical areas.
** Only allocated for broadcasting in Europe and Asia.

viable to operate as many transmitters. Coupled with this it is just as important to ensure that the population is able to pick up radio broadcasts to keep them aware of the news and national events. Having a radio station also helps to maintain a sense of national identity.

The solution is to use frequencies that enable a greater coverage to be gained. As a result a number of 'tropical bands' are allocated with frequencies up to about 5 MHz. By using the relatively short skip normally present on these bands greater coverage can be achieved. Often transmitters in the medium wave and VHF FM are used for the populated areas while the tropical bands are used to give greater coverage for the outlying areas.

The tropical bands are only used by countries between latitudes of 23°N and 23°S. This area covers part of Africa, Asia, Central and South America.

With the large amount of pressure on the short wave broadcast allocations it is necessary for broadcasters to make the maximum use of the available spectrum. Although the channel spacing is only 5 kHz instead of the greater spacing used on the medium wave band spectrum availability is still tight. To improve the spectrum usage and reduce congestion single sideband transmissions are starting to be used and a gradual changeover is being made. When broadcast stations use SSB, they only transmit one sideband. Also the level of the carrier is reduced by 6 dB which enables major savings to be made in the power required for the transmitter.

VHF FM

One of the main disadvantages of AM is the fact that the transmissions are subject to interference and noise. The bandwidths used also mean that the audio frequency response is limited. The VHF FM broadcasts offer much higher quality both in terms of the noise performance and frequency response. With a deviation of ± 75 kHz the transmissions have an overall transmission bandwidth of 200 kHz. The upper audio frequency limit is generally taken as 15 kHz for these transmissions. This is quite adequate for most high quality transmissions.

One of the problems with these high quality VHF FM transmissions is that the increased audio bandwidth means that noise can often be perceived. Even then it is considerably better than that obtained using an AM system. It is particularly noticeable towards the treble end of the audio spectrum, where it can be heard as a background hiss. To overcome this it is possible to increase the level of the treble frequencies at the transmitter. At the receiver they are correspondingly attenuated to restore the balance. This also has the effect of reducing the treble background hiss that is generated in the receiver. The process of increasing the treble signals is called pre-emphasis, and reducing them in the receiver is called de-emphasis. The rate of pre-emphasis and de-emphasis is expressed as a time constant. It is the time constant of the capacitor–resistor network used to give the required level of change. In Europe and Australia the time constant is 50 µs whereas in North America it is 75 µs. Slight modifications are required for the circuitry for receivers supplied to different parts of the world, although it is perfectly possible to receive stations with passable quality even when the time constants are incorrect. The difference would be easily noticed on a hi-fi receiver, although on a small portable radio the difference might not be very noticeable.

Stereo

In line with the improved quality available on the VHF FM broadcasts, it is possible to transmit stereo. However, this has to be accomplished in such a way that ordinary mono radios can still receive the transmissions without any degradation in performance.

A stereo signal consists of two channels that can be labelled L and R (Left and Right), one channel for each speaker. The ordinary mono signal

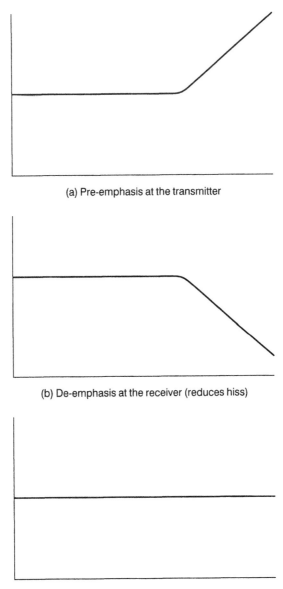

(a) Pre-emphasis at the transmitter

(b) De-emphasis at the receiver (reduces hiss)

(c) Overall response from transmitter input to receiver output

Figure 7.1   *Pre-emphasis and de-emphasis of a signal*

consists of the summation of the two channels, i.e. L + R, and this can be transmitted in the normal way. If a signal containing the difference between the left and right channels, i.e. L – R, is transmitted then it is possible to reconstitute the left and right only signals. By adding the sum and difference signals, i.e. (L + R) + (L − R), gives 2L, i.e. the left signal, and subtracting the two signals, i.e. (L + R) − (L − R), gives 2R, i.e. the right signal. This can be achieved relatively simply by adding and subtracting the two signals electronically. It only remains to find a method of transmitting the stereo difference signal in a way that does not affect any mono receivers.

This is achieved by transmitting the difference signal above the audio range. It is amplitude modulated onto a 38 kHz subcarrier. Both the upper and lower sidebands are retained, but the 38 kHz subcarrier itself is suppressed to give a double sideband signal above the normal audio bandwidth as shown in Figure 7.2. This whole baseband is used to

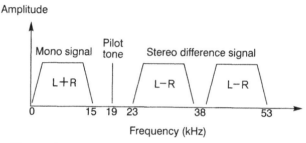

Figure 7.2 *The modulating (baseband) signal for a stereo VHF FM transmission*

frequency modulate the final radio frequency carrier. This signal is also what is regenerated after the signal is demodulated in the receiver.

To regenerate the 38 kHz subcarrier, a 19 kHz pilot tone is transmitted. This provides a reference in the receiver when the signal is reconstituted. The frequency of the 19 kHz pilot tone is doubled in the receiver to give the required 38 kHz signal. This is used to demodulate the signal by reinserting the subcarrier into the double sideband stereo difference signal.

The presence of the pilot tone is also used to detect whether a stereo signal is being transmitted. If it is not present the stereo reconstituting circuitry is turned off. As higher noise levels are present when stereo is being received it enables the stereo circuitry to be on only when it is required. Also when the signal is weak and there is a high background noise level, the pilot tone will not be detected and the stereo circuitry can be switched out of circuit.

To generate the stereo signal, a system similar to that shown in Figure 7.3 is used. The left and right signals enter the encoder where they are passed through a circuit to add the required pre-emphasis. After this they are passed into a matrix circuit. This adds and subtracts the two signals

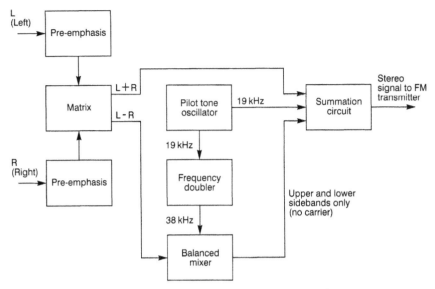

Figure 7.3   *A simplified diagram of a VHF FM stereo encoder*

to provide the L + R and L – R signals. The L + R signal is passed straight into the final summation circuit to be transmitted as the ordinary mono audio. The difference L – R signal is passed into a balanced modulator to give the double sideband suppressed carrier signal centred on 38 kHz. This is passed into the final summation circuit as the stereo difference signal. The other signal entering the balanced modulator is a 38 kHz signal that has been obtained by doubling the frequency of the 19 kHz pilot tone. The pilot tone itself is also passed into the final summation circuit. The final modulating signal consisting of the L + R mono signal, 19 kHz pilot tone and the L – R difference signal based around 38 kHz is then used to frequency modulate the radio frequency carrier before being transmitted.

Reception of a stereo signal is very much the reverse of the transmission. A mono radio receiving a stereo transmission will only respond to the L + R signal. The other components being above 15 kHz are above the audio range, and in any case they will be suppressed by the de-emphasis circuitry.

A variety of methods are available for decoding the stereo signal once it has been demodulated in the radio. The basic method and circuitry for demodulating an FM signal are described in Chapter 5 (Receivers). This produces the baseband signal consisting of the L + R, L – R and the pilot tone. These have to be extracted and converted into the two audio signals L and R.

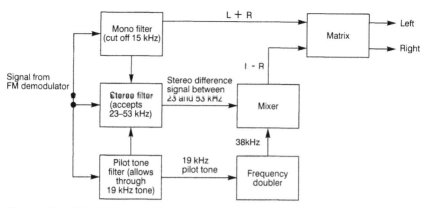

Figure 7.4 *Block diagram of a stereo decoder*

The block diagram of one type of decoder is shown in Figure 7.4. Although this is not the only method which can be used it shows the basic processes that are required. The signal is first separated into its three constituents. The L + R mono signal between 0 and 15 kHz, the pilot tone at 19 kHz and the stereo difference signal situated between 23 and 53 kHz. First, the pilot tone at 19 kHz is doubled in frequency to 38 kHz. It is then fed into a mixer with the stereo difference signal to give the L – R signal at audio frequencies. Once the L + R and L – R signals are available they enter a matrix where they are added and subtracted to regenerate the L and R signals. At this point both signals are amplified separately in the normal way in a stereo amplifier before being converted into sound by loudspeakers or headphones.

Today most stereo radios use an integrated circuit to perform the stereo decoding. Often the pilot tone is extracted and its frequency is doubled using a phase locked loop. This provides a very easy and efficient method of performing this function without the need for sharp filters.

RDS

The initials RDS stand for radio data system. It is a system that uses inaudible data signals added to a VHF FM transmission to bring a variety of information and automatic tuning facilities to the listener. Initially aimed towards the car radio market RDS is now a standard feature on many hi-fi tuners and other sets. The system is well established in Europe and is used in a number of other countries as well.

The concept of using a subcarrier included in the baseband modulation to carry data for information and other services dates back to the 1960s. In the USA many FM stations use a subcarrier to carry additional low bandwidth subsidiary audio for use as background music in areas such as shops or stores. In view of this it is called 'storecasting'. This is covered by the Subsidiary Communications Authorization of the FCC, and as a result the system is also referred to as SCA. In Europe this idea was not taken up as it was felt that the level of crosstalk from the SCA channel to the main programme was too high. Instead it was felt that a low data rate channel would be able to operate without giving rise to unacceptable levels of crosstalk. Initial development of various ideas started in the early 1970s but it was not until 1976 that work was co-ordinated across Europe and in 1983 the final specification was agreed.

The system operates by adding a data signal to the baseband modulation. It is carried above the stereo difference signal on a 57 kHz subcarrier. This is three times the stereo pilot carrier as shown in Figure 7.5. This was chosen because the data is based around a carrier at a

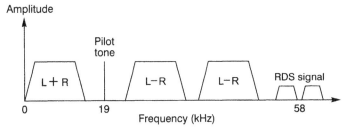

Figure 7.5   *Baseband modulation for an FM signal carrying RDS*

harmonic of the pilot tone that is also used for the stereo signal and as a result it creates fewer audible intermodulation products. The data is modulated onto this subcarrier using a form of modulation called quadrature phase shift keying (QPSK). This gives good immunity to data errors while still allowing sufficient data to be carried. Being a phase shift keying system it also gives good immunity to interference to the audio signal being carried.

The data message is divided into small separate entities that can be received, decoded and processed independently to overcome the effects of signal variation and multipath distortion. Data is transmitted at a rate of 1187.5 bits per second which is equal to the RDS subcarrier frequency divided by 48. By adopting this approach the decoding circuits are able to operate synchronously, for improved performance.

Data is transmitted in groups consisting of four blocks, each containing a 16-bit information word and a 10-bit check word as shown in Figure 7.6. The large check word is required to ensure the system operates correctly even under conditions of low signal strength or high interference. Once

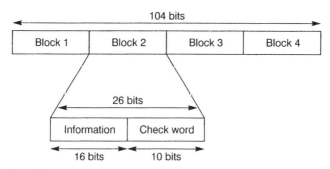

Figure 7.6 *RDS data structure*

detected it also allows the receiver to correct errors, thereby making the system far more resilient.

Once generated the data is encoded onto the subcarrier in a differential format. When the data level is at a logical '0' the output remains unchanged. When a logical '1' appears the output changes state. The baseband signal has to have its spectrum carefully limited. This is done to ensure that no crosstalk between any of the decoders takes place. This is achieved by encoding each bit as a biphase signal and also by passing the coded signal through a filter.

RDS provides a variety of facilities not previously available with VHF FM broadcasts. The most widely publicized is the travel service that is of particular use for car radios. There are two flags associated with travel messages. The first is the TP flag. This indicates that a station carries travel announcements. The second is the TA flag that indicates when a travel message is actually being transmitted. When the TA flag is asserted the receiver may pause the tape, turns up the volume or takes the receiver out of a muted condition so that the message can be heard. After the message is complete the TA flag returns to normal and the receiver resumes its previous state.

RDS also enables tuning facilities to be improved. Again this is very useful for car radios. As the car moves from the service area of one transmitter to the next it can automatically tune to the strongest transmitter. The set accomplishes this by looking at the alternative frequency (AF) list and checks that signals on these channels have the correct programme identity (PI) code. The AF codes carry the VHF Band II channels on which the station may be heard. These channels start at 87.5 MHz with channel 0, then 87.6 is channel 1, and so forth. Up to 25 alternative channels may be catered for under the basic system, although this can be expanded to cater for more when required. Normally the list is kept as short as possible, but large lists may be needed when there are small relay stations that do not decode the data, and only retransmit the full complete signal. When radios fitted with RDS store the frequency of a station, they normally store the programme identification (PI) code along side it. In this way if the receiver is located outside the coverage area for the given transmitter it can still locate the correct network.

While the system operates in this manner for national networks some changes have to be made for local stations. Here the radio will tune to the strongest station of the same type when moving out of the service area of the first station.

The PI code is structured, consisting of four characters. The first indicates the country of origin, and it is 'C' for the UK. The second indicates the type of coverage. The figure '2' indicates a national station, and the final two letters are used for the programme reference. As an example C203 is the code for the BBC Radio 3 network and C311 is used for the BBC local radio station GLR.

It can be imagined that when moving from the service area of one transmitter to the next there may be a small gap as the receiver checks for the strongest signal, and that it has the correct PI. This can be improved by employing a second front end section in the set to check the alternative frequencies. Some sets incorporate this and the change from one transmitter to the next is virtually unnoticeable.

Another facility which is becoming increasingly useful is the programme service (PS) name. This enables the receiver to display the station name, a facility which is very useful with the large variety of stations on the air today.

A further facility which was not available at the launch of RDS is called enhanced other networks (EON). This allows the set to listen to a national network, but to be interrupted by travel news on a local station. This enables the set to be sensitive only to relatively local traffic problems even though the set is tuned to a national network. To achieve this facility requires a large amount of co-ordination between the different stations. To implement this feature the BBC have a central computer specifically for this. When a local station is about to transmit a traffic message this fact is communicated to this computer. It directs the relevant transmitters in the national network to indicate this fact, enabling the receiver to retune to the relevant local radio station. The data transmitted includes an alternative frequencies list that may be used to pick up the required local station, and on the best frequency for the particular location of the receiver.

To be able to receive RDS data three circuit blocks are required. The first is a 57 kHz filter which is used to isolate the signals from the baseband audio and the stereo difference signals. Next this must be fed into a demodulator. This removes the data from the 57 kHz subcarrier. Once the data is available it needs to be processed for use by the receiver. As most car radio sets these days incorporate a processor, it can also be used to perform this function.

Table 7.2 *RDS terminology*

AF	Alternative frequencies – A list of the frequencies used by a station in adjacent coverage areas.
CT	Clock time and date – Data containing time and date information. This enables the clock to display the correct time and adjust between changes in winter and summer time without the need for manual setting.
DI	Decoder information – This signal allows for miscellaneous functions to be controlled in the radio.
EON	Enhanced other networks – Information which is transmitted giving the radio a cross-reference to other stations when travel information is being transmitted.
MS	Music/Speech – A data flag that allows for the relative levels of speech and music to be altered.
PI	Programme identification – This is a station code used in conjunction with the alternative frequency signal to locate suitable alternative transmitters for a given programme. Each service is allocated its own unique identifier.
PIN	Programme identification number – This signal identifies a given programme, and allows for a radio or recorder to be turned on when the identification number is recognized.
PTY	Programme type selection – This signal identifies one of 15 types of programme, allowing selection of the type of listening rather than by station.
PS	Programme service name – A signal that carries the station name and allows this to be displayed by the receiver.
RT	Radio text – This allows information about the programme to be displayed by the radio.
TDC	Transparent data channel – This allows data to be downloaded over the radio.
TP/TA	Travel service – These signals enable the travel information to be heard, regardless of the choice of listening.

Digital radio (digital audio broadcasting – DAB)

Most current broadcast techniques use analogue techniques to carry the audio on the carrier. However, with the move to the use of digital techniques because of the improved performance a digital system is now available in a number of countries. Called digital audio broadcasting

(DAB) it gives many improvements and new features over the existing AM and FM systems in use.

DAB gives major improvements over FM, particularly for those in automobiles. FM suffers particularly when signal strengths are low and the signal is being received by a number of different paths. Reflections off buildings, hills and other objects cause distortion of the signal, and in addition to this the tuning has to be altered when passing from the coverage area of one transmitter to the next. With the increase in the number of facilities being expected, new data capabilities are also required. All of these have been incorporated into DAB together with an improved ease of tuning.

The problem of reflections is at the core of determining how the new system may work. Although digital audio is widely used in television broadcasts, these have the advantage that direction antennas are used which will greatly limit the effect of any reflections. For a car system that uses an omnidirectional antenna, any system must be resilient to reflections, without suffering any degradation in performance. The system is also designed to make tuning easier. On the medium wave or VHF FM bands, there are many stations that can be heard, often making it difficult to locate the wanted station. DAB aims to simplify this to a very large degree.

The main problem with a digital system is to ensure that reflections do not corrupt the data being sent as shown in Figure 7.7. For satisfactory operation the system must be able to cope with significant delays. This reduces the data rate well below that required for a system using normal linear encoding like that used for a CD. For these discs the data rate is about 1.4 Mbits per second.

To enable the data to be successfully transmitted, two data rate reduction systems are employed. The first is an audio processing system, and the second is a totally different form of modulation called coded orthogonal frequency division multiplex (COFDM).

The audio processing system relies upon the fact that the ear can only detect certain sounds. Below a given level it does not hear them. This

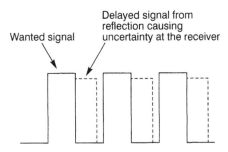

Figure 7.7 *Data corruption caused by reflections*

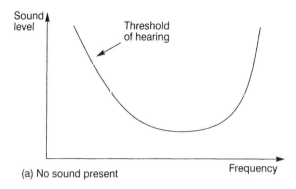

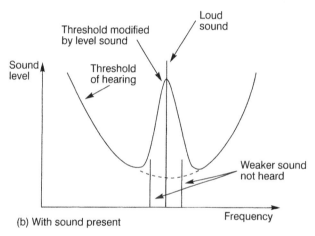

Figure 7.8   *Threshold of hearing of the ear*

threshold of hearing varies with frequency as shown in Figure 7.8. The presence of sounds also varies the threshold. A strong signal in one portion of the audio spectrum will mask weaker ones close to it as shown. By detecting the threshold of hearing and not encoding signals below that level the data rate can be significantly reduced, in this system by a factor of between six and 12 when compared to a linearly encoded system.

To extract the perceptible part of the audio, the spectrum is split into 32 equally spaced bands. The signal in each of these bands is quantized in such a way that the quantizing noise from the digitizing process matches the masking threshold.

The sampling frequency is generally 48 kHz although 24 kHz can be used for lower quality transmissions. Different encoded bit rate options are available. These are 32, 48, 56, 64, 80, 96, 112, 128, 160 and 192 kbits per second for each monophonic channel. There is a stereo mode, or for low bit rate transmissions a mode known as joint stereo mode may be used. This uses the redundancy and interleaving of the two channels of a stereophonic programme to maximize the overall perceived audio quality.

The modulation system uses a large number of low data rate carriers packed closely together. This is achieved by using frequency division mutliplexing, and overlapping the signals slightly. However, by altering the phase so that they are orthogonal to one another the individual carriers can be demodulated without any interference from one another.

A total of just over 1500 carriers occupying a bandwidth of 1.5 MHz (1.536 MHz to be exact) are used to carry the data, and they have a spectrum like that shown in Figure 7.9. A guard band of 250 kHz is placed

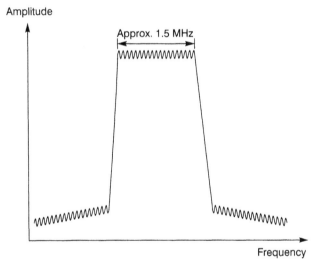

Figure 7.9 *Spectrum of a DAB signal*

between adjacent channels to prevent interference between different transmissions. This allows data at a rate of just over 1.5 Mbits per second to be carried giving the capability of up to six high quality stereo transmissions to be carried or 20 or so low quality mono services. It is possible to alter the data rate according to the requirements of the programme so that maximum use is made of the available capacity. Additional data services are also available.

The transmitted data is used for a variety of different services. The primary one is obviously the audio. This can be apportioned as required and can be selected to be between 32 and 384 kbits per second, allowing typically up to six high quality stereo transmissions or up to 20 restricted quality mono programmes, or a mixture as required.

Programme associated data (PAD) is embedded in the audio bit stream and is used for data transmitted in support of the audio programme. It may include information about the track in the form of the artist's name, the title of the track, lyrics, or even graphics for display on the receiver. The data may also be used for facilities such as dynamic range control. Like the audio data rate, the programme associated data rate is adjustable, but with a minimum data rate of 667 bits per second and a maximum of 65 kbits per second.

In addition to the PAD, general data may be transmitted as a separate service. This can be achieved in a number of ways. It may be via what is termed the fast information channel (FIC) or it may be as packets of data in a packet sub-multiplex. This service may be used for a variety of purposes from a traffic message channel to an electronics newspaper.

Data may also be transmitted to support pay to listen services. Conditional access (CA) data may be transmitted, although the format for this is not defined yet. This is expected to be determined by the subscription service using the facility.

Further data in the form of service information (SI) is transmitted. This is used for the control of receivers and to provide information for programme selection to the user. It can also be used to provide links between different services in the multiplex. It can even be used to provide links to services in other DAB multiplexes or ensembles or even to AM or FM broadcasts.

The DAB transmissions are very efficient in their use of spectrum. In view of their resilience to interference it means that adjacent transmitters do not need to use different frequencies as is the case for normal AM and FM services. Instead the same frequency is reused as required to give continuous coverage in what is termed a single frequency network (SFN). In the UK, a single national network for FM occupies 2.2 MHz. A DAB transmission occupying only 1.5 MHz can carry several audio services as well as having an additional data capability.

To enable the receiver to be able to decode the data being transmitted it must have a defined format. This is shown in Figure 7.10. Each

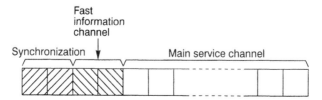

Figure 7.10 *DAB data frame structure*

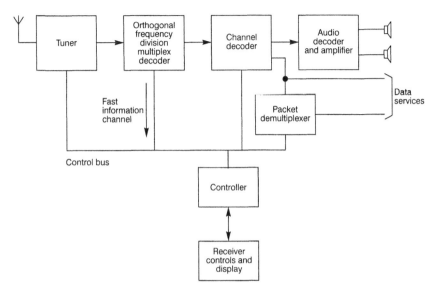

Figure 7.11   *Block diagram of a DAB receiver*

transmission frame starts with a null signal to give approximate synchronization. This is followed by a phase reference for the demodulation process. After these references the next symbols of data are reserved for the fast information channel (FIC) used for data. The remaining frames are then used for the main service channel (MSC). Each programme or service within the MSC section of the frame is allocated its own time for decoding purposes. The total length of the frame for terrestrial transmissions which are being made at the moment is 96 ms, although additional standards exist for 48 or 24 ms. These are intended for use on other frequencies and for satellite broadcasts which are planned for the future.

Receivers for DAB naturally require totally different decoding circuitry to that used for more typical AM or FM receivers. They make heavy use of digital signal processing techniques to reconstruct and process the data. A typical receiver block diagram is shown in Figure 7.11. Circuitry up to the intermediate frequency stages is similar to that of a normal radio receiver, using the superhet principle. Even so the IF stage must be sufficiently wide to accommodate the 1.5 MHz signal. After the IF stages the similarities end because the signal enters an orthogonal frequency division multiplex decoder. This reconstitutes the raw data which is further processed as shown to produce the audio for amplification, or data for display or other uses, including being linked into PCs, etc.

8 Satellites

Satellites have become an accepted part of today's technology. They are used for a variety of purposes from providing links for long-distance telecommunications, to direct broadcast television and radio, gathering weather and geological information, as well as a host of other facilities. To reach where we are today has taken a tremendous amount of development and investment since the idea of communications satellites was first proposed by Arthur C. Clarke in an article in the British magazine entitled *Wireless World* in 1945. Even before this, Sir Isaac Newton proposed the ideas of satellites, sketching ideas for placing an 'artificial moon' in orbit by firing it from a canon placed on the top of a mountain. It was not until 4 October 1957 that the first satellite was launched. Named Sputnik, it weighed just over 80 kg and was placed into a low earth orbit, circling the earth every 96 minutes. This meant that despite its relatively small size people could actually see it on earth with the naked eye.

Since Sputnik the development of satellites has continued at a swift pace. It did not take long before other satellites followed, each one enabling the technology to develop a little further. The electronics to enable all the functions in today's highly sophisticated satellites to be performed has been developed along with the launch vehicles capable of putting these satellites, often as heavy as a ton, into space. Both of these elements of the development have required a large investment. At the time of Arthur C. Clarke's article the most advanced rockets were the V2s, developed by the Germans, and these had never been intended to launch satellites into outer space. For the satellites themselves, methods of powering them had to be developed along with control systems to enable them to be finely positioned. The satellites also had to be made very reliable, as they cannot be serviced once they have been launched. Nowadays all these problems have been mastered and satellites are capable of providing a long reliable service life.

Satellite orbits

There are a variety of different orbits that can be adopted for satellites. The one that is chosen will depend on factors including its function, and the area it is to serve. In some instances the orbit may be as low as 100 miles (160 km), whereas others may be over 22 000 miles (36 000 km) high.

As satellites orbit the earth they are pulled back in by the force of the gravitational field. If they did not have any motion of their own they would fall back to earth, burning up in the upper reaches of the atmosphere. Instead the motion of the satellite rotating around the earth has a force associated with it pushing it away from the earth. For any given orbit there is a speed for which these two forces balance one another and the satellite remains in the same orbit, neither gaining height nor losing it.

Obviously the lower the orbit, the stronger the gravitational pull, and this means that the satellite must orbit the earth faster to counteract this pull. Further away the gravitational field is less and the satellite velocities are correspondingly less. For a very low orbit of around 100 miles a velocity of about 17 500 miles per hour is needed and this means that the satellite will orbit the earth in about 90 minutes. At an altitude of 22 000

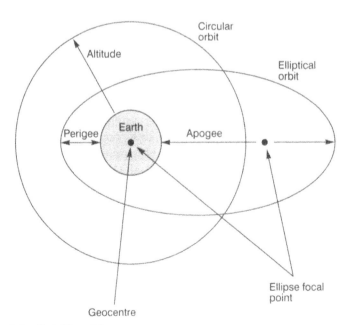

Figure 8.1 *Satellite orbits*

miles a velocity of just less than 7000 miles per hour is needed giving an orbit time of about 24 hours.

A satellite can orbit the earth in one of two basic types of orbit. The most obvious is a circular orbit where the distance from the earth remains the same at all times. A second type of orbit is an elliptical one. Both types of orbit are shown in Figure 8.1, where the main characteristics are shown.

When a satellite orbits the earth, either in a circular or elliptical orbit it forms a plane. This passes through the centre of gravity of the earth or the geocentre. The rotation around the earth is also categorized. It may be in the same direction as the earth's rotation when it is said to be posigrade, or it may be in the opposite direction when it is retrograde.

At any given time there is a point on the earth at which the satellite is directly overhead. As the satellite moves so does this point, and the track that this traces out on the earth's surface is known as the groundtrack. The groundtrack is a great circle, i.e. the centre of the circle is at the geocentre. Geostationary satellites are a special case as they appear directly over the same point of the earth all the time and accordingly their groundtrack consists of a single point on the earth's equator. For satellites with equatorial orbits the groundtrack is along the equator.

Satellites may also be in other orbits. These will cross the equator twice, once in a northerly direction and once in a southerly direction. The point at which the groundtrack crosses the equator is known as a node. There are two, and the one where the groundtrack passes from the southern hemisphere to the northern hemisphere is called the ascending node. The one where the groundtrack passes from the northern to the southern hemisphere is called the descending node. For these orbits it is usually found that the groundtrack shifts towards the west for each orbit because the earth is rotating towards the east underneath the satellite.

For many orbit calculations it is necessary to consider the height of the satellite above the geocentre. This is the height above the earth plus the radius of the earth. This is generally taken to be 3960 miles or 6370 km.

Velocity is another important factor as already seen. For a circular orbit it is always the same. However, in the case of an elliptical one this is not the case as the speed changes dependent upon the position in the orbit. It reaches a maximum when it is closest to the earth and it has to combat the greatest gravitational pull, and it is at its lowest speed when it is furthest away.

For an elliptical orbit the centre of the earth forms one of the focal points of the ellipse as shown in Figure 8.1 (see above). It can also be seen that there are points where the satellite is furthest from the earth, and closest to it. These are important and they are called the apogee (furthest away) and perigee (closest). Generally the altitudes at the apogee and perigee are calculated from the geocentre.

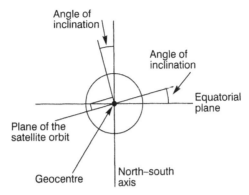

Figure 8.2   *Angle of inclination of an orbit*

A satellite may orbit around the earth in different planes. The angle of inclination of a satellite orbit is shown in Figure 8.2. It is the angle between a line perpendicular to the plane of the orbit and a line passing through the poles. This means that an orbit directly above the equator will have an inclination of 0 degrees (or 180 degrees), and one passing over the poles will have an angle of 90 degrees. Those orbits above the equator are generally called equatorial obits, while those above the poles are called polar orbits.

Another important factor about the position of a satellite is its angle of elevation. This is important because the earth station will only be able to maintain contact with the satellite when it is visible as only direct or line of sight communications are possible with the satellite. The angle of elevation is the angle at which the satellite appears above the horizontal. If the angle is too small then signals may be obstructed by nearby objects if the antenna is not very high. For those antennas that have an unobstructed view there are still problems with small angles of elevation. The reason is that signals have to travel through more of the earth's atmosphere and are subjected to higher levels of attenuation as a result. An angle of 5 degrees is generally accepted as the minimum angle for satisfactory operation.

In order that a satellite can be used for communications purposes the ground station must be able to follow it in order to receive its signal, and transmit back to it. Communications will naturally only be possible when it is visible, and dependent upon the orbit it may only be visible for a short period of time. To ensure that communication is possible for the maximum amount of time there are a number of options that can be employed. The first is to use an elliptical orbit where the apogee is above the planned earth station so that the satellite remains visible for the maximum amount of time. Another option is to launch a number of satellites with the same orbit so that when one disappears from view, and communications are lost, another one appears. Generally three satellites are required to maintain almost uninterrupted communication. However, the handover from one satellite to the next introduces additional complexity into the system, as well as having a requirement for three satellites.

The most popular option is to use a satellite in what is called geostationary orbit. Using this orbit the satellite rotates in the same direction as the rotation of the earth and has a 24 hour period. In this way it revolves at the same angular velocity as the earth and in the same direction. As a result it remains in the same position relative to the earth. For the relative forces to balance and the satellite to remain a geostationary orbit it must be above the equator and have an altitude of 22 300 miles or 35 860 km. Geostationary orbits are very popular because once the earth station is set onto the satellite it can remain in the same position, and no tracking is normally necessary. This considerably simplifies the design and construction of the antenna. For direct broadcast satellites it means that people with dishes outside the home do not need to adjust them once they have been directed towards the satellite.

The path length to any geostationary satellite is a minimum of 22 300 miles. This gives a small but significant delay of 0.24 seconds. For a communications satellite this must be doubled to account for the uplink and downlink times giving virtually half a second. This delay can make telephone conversations rather difficult when satellite links are used. It can also be seen when news reporters are using satellite links. When asked a question from the broadcasters studio, the reporter appears to take some time to answer. This delay is the reason why many long-distance links use cables rather than satellites as the delays incurred are far less.

Satellites

The requirements for satellites are very stringent, both in terms of their design and construction. They must be capable of operating in extreme conditions while still maintaining the highest standards of reliability because they cannot be retrieved for maintenance or repair. Satellites also contain a number of systems used for what is called station keeping.

These systems are used to maintain the satellite in the correct orbit and position. The satellite will tend to drift away from its correct position over a period of time. To correct this small thrusters are used. Often they consist of canisters of a gas which when released with a catalyst gives a form of rocket propulsion to move the satellite back on station. Often the service life of a satellite is determined by the amount of fuel for repositioning the satellite rather than the reliability of the electronics.

The other problem with a satellite is that its attitude will change. This is of great importance because directive antennas or cameras are often used, and the satellite needs to be orientated in the correct direction for them. The basic method of gaining the correct orientation is to use the thrusters. However, the attitude will change comparatively quickly. The most common method to overcome this is to use the gyroscopic effect. Sometimes a large flywheel may be made to spin inside the satellite. This can be inefficient in its use of the weight of the satellite. To overcome this other cylindrical satellites actually rotate a portion of the body, often an inner cylindrical section so that the antennas mounted on the outer section do not revolve.

Electrical power is also required by the satellite for its electronic circuitry and other electrical systems. This is supplied by the large arrays of photo or solar cells that are used. Some cylindrical satellites have them positioned around the outer area on the cylinder so that some part of the body is always exposed to sunlight. Others have large extending panels that are orientated to collect the maximum amount of light. Today these

Figure 8.3 *A typical satellite (courtesy NASA/JPL/Caltech)*

panels are capable of producing the many kilowatts of power required for the high power output stages used in many transponders.

The satellite also needs batteries that charge up when the satellite is in sunlight. These need to charge sufficiently from the solar cells so that when the satellite passes out of the sunlight it can still remain operational. They should be sufficient to power the satellite for the full period of darkness. This naturally places an additional burden on the solar cells because they need to be able to power not only the satellite itself, but also charge the batteries. This can double the power they have to supply during periods of sunlight.

Satellites must also be designed and manufactured to withstand the harsh environment encountered in outer space. Extremes of temperature are encountered. The surfaces exposed to the sun are heated by solar radiation and will rise to very high temperatures, whereas the other side that is not heated will be exceedingly cold. Only conduction will give any heating effect under these circumstances.

There are a number of other effects that must be considered. Solar radiation itself has an effect on some materials, causing them to degrade. Notice must also be taken of meteorites. Very small ones cause the surfaces to be eroded slightly, but larger ones may penetrate the body of the satellite causing significant damage. To overcome this satellites are protected by specially designed outer layers. These consist of sheets of metal which are slightly separated giving a cushioning effect when any meteorites impact on the satellite. Cosmic particles also degrade the performance of satellites. Particularly during solar flares the increase in solar particle flow can degrade solar cells, reducing their efficiency.

The antennas used on satellites are particularly important. For geostationary satellites directional antennas are generally used. These are used because power use has to be optimized and antennas that give gain enable the best use to be made of the available transmitted power. Additionally they enable the signals from the earth to be received with the best signal to noise ratio. In view of the long path lengths required for geostationary satellites, there is a considerable path loss and the antenna gain is used to improve the received signal strength. It also helps reduce the reception of solar and cosmic noise that would further degrade the received signal. In a geostationary orbit the earth subtends only 18 degrees of arc. Any power not falling into this area is wasted.

As a result, parabolic reflector or 'dish' antennas are widely used. Horn antennas are also popular and in some cases phased arrays may be employed, especially where coverage of a specific area of the globe is required. However, the use of directional antennas does mean that the orientation of the antenna is crucial, and any perturbation of the alignment of the satellite can have a major effect on its operation, both in reception and transmission.

For low earth orbit satellites the situation is somewhat different. The fact that they move across the sky and may need to be received by several users at any time means that they cannot use directive antennas. Additionally the earth subtends around half the celestial sphere and as a result users may be separated by angles ranging from zero to almost 180 degrees. Fortunately the satellites are much closer to the earth and path losses are very significantly less, reducing the need to high gain antennas.

Ground stations also need an effective antenna system. For communication with satellites in geostationary orbit the antenna remains fixed, except if there is a need to change to another satellite. Accordingly parabolic reflectors are often used. This can be seen from the number of satellite TV antennas that are in use. These are a form of parabolic reflector. However, it is possible to use other types such as arrays of Yagi antennas. Here they are stacked (one on top of the other) and bayed (side by side) to give additional gain.

For some low earth orbit satellites the ground station antenna systems are designed to be able to track the satellite in azimuth and elevation. This is typically achieved by automatically tracking the satellite as it moves across the sky. This is normally achieved by taking a signal level output from the receiver. By ensuring that it is maintained at its peak level the satellite will be tracked. Many low earth orbit satellites are required for systems such as positioning or even telephone style communications. Here directional antennas are not used as the user will not want to reorientate the antenna all the time. Instead almost non-directional antennas are used and the transmitter powers and receiver sensitivities designed to give a sufficient level of signal to noise ratio.

Placing a satellite in orbit

In view of the colossal amount of energy required to place a satellite in orbit it is necessary to ensure that the energy is used in the most effective way. This ensures that the amount of fuel required is kept to a minimum; an important factor on its own because the fuel itself has to be transported until it is used.

Many satellites are placed into geostationary orbit, and one common method of achieving this is based on the Hohmann transfer principle. This is the method used when the Shuttle launches satellites into orbit. Using this system the satellite is placed into a low earth orbit with an altitude of around 180 miles. Once in the correct position in this orbit rockets are fired to put the satellite into an elliptical orbit with the perigee at the low earth orbit and the apogee at the geostationary orbit as shown in Figure 8.4. When the satellite reaches the final altitude the rocket or

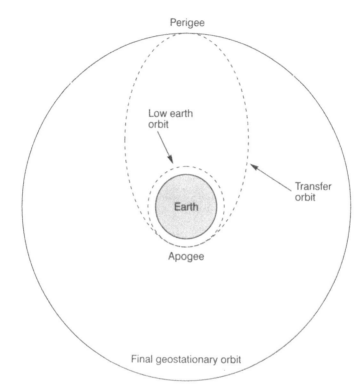

Figure 8.4 *Use of a transfer orbit to place a satellite in geostationary orbit*

booster is again fired to retain it in the geostationary orbit with the correct velocity.

Alternatively when launch vehicles like Ariane are used the satellite is launched directly into the elliptical transfer orbit. Again when the satellite is at the required altitude the rockets are fired to transfer it into the required orbit with the correct velocity.

These are the two main methods of placing satellites into orbit. Naturally it would be possible to place a satellite directly into geostationary orbit, but this would take more energy and would not be feasible.

Path calculations

Communication with satellites occurs via a direct line of sight path. As a result it is possible to calculate the signal levels which will be received by the satellite or the earth station assuming that figures like the transmitter power, antenna gains and receiver performance are known. To ensure that the signal levels are kept within working limits without overdesigning a system that would cost extra money, link budgets are always calculated.

Figure 8.5   *Delta Rocket launch. This launch shows a Delta Rocket carrying the Mars Pathfinder being launched from Cape Canaveral Spaceflight Center at 1:56 am on 4 December 1996 (courtesy NASA/JPL/Caltech)*

The major calculation which is performed when investigating the link budget is the path loss. As the signal travels along a direct line of sight the signal attenuation which is experienced is the free space path loss. This is expressed in decibels as shown below:

$$\text{Loss (dB)} = 22 + 20 \log_{10} (R/\lambda)$$

where R is the range or distance

λ is the wavelength expressed in the same units as the distance

From this it can be seen that the loss increases with frequency, but it should be remembered that at higher frequencies it is possible to use higher gain antennas and this compensates for this. This means that for a satellite in geostationary orbit the loss will be between 195 and 213 dB assuming operating frequencies between 4 and 30 GHz. Although the altitude of the satellite is 22 300 miles, the actual distance will often be greater than this because the earth station will not be directly below the satellite.

Signal effects

While satellites use frequencies that are well above those that are normally thought to be affected by the ionosphere, the ionosphere and indeed other areas of the atmosphere still have an affect. Even though the signals are not reflected by the ionosphere, like those in the HF portion of the spectrum, some effects are still noticed.

One of the most common is an effect known as Faraday rotation. This effect is more pronounced at lower frequencies, and at frequencies above 10 GHz it is seldom noticed. However, the effect is seen on linearly polarized signals where the polarization of a linearly polarized signal is seen to randomly change with time. If a signal is circularly polarized then the rotation has no effect.

Tropospheric bending can also affect signals. This occurs particularly when the satellite is low in the sky and the path length through the troposphere is longer. When this occurs the path losses may increase. Again the effect is less pronounced at higher frequencies.

The Doppler effect can also be noticed under some circumstances. Naturally this is not noticed in the case of geostationary satellites because they are stationary above a given location. However, it is often noticed when using satellites such as those in low earth orbits or elliptical orbits that move relative to a point on the earth's surface. Relative velocities may be quite high, and frequencies are often as high as several GHz resulting in some noticeable frequency shifts. As a result some satellite systems have to be able to accommodate frequency shifts arising from this.

Communications satellites

Satellites fulfil a number of roles. They can be used for communications, direct broadcasting, weather monitoring, navigation and a number of other uses. Of these the most widespread is possibly their use in communications. Every day many thousands of international telephone calls are made. These are often carried via satellites. In the early days of telephones, calls were routed via undersea cables. These were expensive to lay, and were difficult to maintain. They also had a relatively narrow bandwidth and this meant the number of calls they could carry was limited.

For use in communications like this the satellite acts as a repeater. Its height above the earth means that signals can be transmitted over distances that are very much greater than the line of sight as shown in Figure 8.6. An earth station transmits the signal, normally consisting of

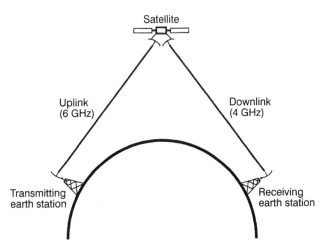

Figure 8.6   *Using a satellite for long distance communications*

many telephone calls up to the satellite. This is called the uplink and might be on a frequency of 6 GHz. The satellite receives the signal and retransmits it. This might be on a frequency of 4 GHz on what is termed the downlink. This has to be on a different frequency to avoid interference between the two signals. If the transmission took place on the same frequency as the received signal, then the transmitted signal would overload the receiver in the satellite, preventing reception on the uplink. The use of satellites is very widespread and a wide variety of frequencies are used. While the figures included here are typical, many other frequencies are also used.

The circuitry in the satellite that acts as the receiver, frequency changer and transmitter is called a transponder. As shown in Figure 8.7, this basically consists of a low noise amplifier, a frequency changer consisting of a mixer and local oscillator, and then a high power amplifier. The filter on the input is used to make sure that any out of band signals such as the transponder output are reduced to acceptable levels so that the amplifier

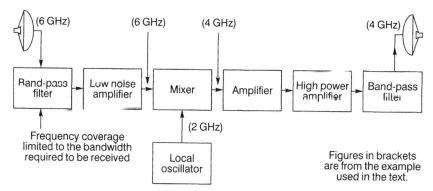

Figure 8.7 *Block diagram of a basic transponder*

is not overloaded. Similarly the output from the amplifiers is filtered to make sure that spurious signals are reduced to acceptable levels. Figures used in here are the same as those mentioned earlier, and are only given as an example. The signal is received and amplified to a suitable level. It is then applied to the mixer to change the frequency in the same way that occurs in a superhet receiver. As a result the satellite receives in one band of frequencies and transmits in another.

In view of the fact that the receiver and transmitter are operating at the same time and in close proximity, care has to be taken in the design of the satellite that the transmitter does not interfere with the receiver. This might result from spurious signals arising from the transmitter, or the receiver may become desensitized by the strong signal being received from the transmitter. The filters already mentioned are used to reduce these effects.

Signals transmitted to satellites usually consist of a large number of signals multiplexed onto a main transmission. In this way one transmission from the ground can carry a large number of telephone circuits or even a number of television signals. This approach is operationally far more effective than having a large number of individual transmitters.

Obviously one satellite will be unable to carry all the traffic across the Atlantic. Further capacity can be achieved using several satellites on different bands, or by physically separating them from one another. In this way the beamwidth of the antenna can be used to distinguish between different satellites. Normally antennas with very high gains are used, and these have very narrow beamwidths, allowing satellites to be separated by just a degree or so.

Another variant of communications satellites is those used for direct broadcasting. This form of broadcasting has become very popular as it provides very high levels of bandwidth because of the high frequencies

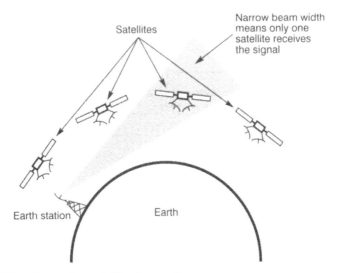

Figure 8.8 *Separating satellites by position*

used. This means that large numbers of channels can be carried. It also enables large areas of the globe to be covered by one delivery system. For terrestrial broadcasting a large number of high power transmitters are required that are located around the country. Even then coverage may not be good in outlying areas.

These DBS satellites are very similar to ordinary communications satellites in concept. Naturally they require high levels of transmitted power because domestic users do not want very large antennas on their houses to be able to receive the signals. This means that very large arrays of solar cells are required along with large batteries to support the broadcasting in periods of darkness. They also have a number of antenna systems accurately directing the transmitted power to the required areas. Different antennas on the same satellite may have totally different footprints.

Satellites have also been used for cellular style communications. They have not been nearly as successful as initially anticipated because of the enormously rapid growth of terrestrial cellular telecommunications, and its spread into far more countries and areas than predicted when the ideas for satellite personal communications was originally envisaged. The systems that were set up used low earth orbiting satellites, typically with a constellation of around 66 satellites. Hand-held phones then communicated directly with the satellites which would then process and relay the signals as required.

The main advantage of the satellite system is that it is truly global and communications can be made from ships, in remote locations where there would be no possibility of there being a communications network. However, against this the network is expensive to run because of the cost of building and maintaining the satellite network, as well as the more sophisticated and higher power handsets required to operate with the satellite. As a result these networks have not been financially successful.

Navigational satellites

One of the most recent developments in satellite applications is in the Global Positioning System or GPS. Although initially conceived as a military system, it is now very widely used for commercial applications. Part of the appeal of the system is that it gives worldwide coverage with accuracies of just a few metres. In recent years the system has come into widespread use. Receivers for domestic use are available at low cost enabling them to be used in domestic vehicles, and in many other everyday applications as well as for very many professional and military uses.

The system uses a constellation of 24 satellites orbiting the earth at an altitude of 20 183 km. Not all the satellites are operational at any given time. Of the total quantity there is a minimum of 21 that are operational 98 per cent of the time to provide the required level of service. The satellites are arranged in six orbital planes of 55 degrees inclination (four satellites per plane) and they take 11 hours 58 minutes (12 sidereal hours) to complete each orbit. This arrangement enables each receiver to pick up at least four satellites under normal conditions. In fact satellites are visible above the horizon for five hours.

Each satellite transmits a message containing three pieces of information: the satellite number, its position in space and the time at which the message was sent. Each satellite is held in a very accurate orbit, and on board it contains an atomic clock to give a highly accurate time source. This time information is transmitted with the signal from each satellite. From this it is possible for the receiver to calculate the different path lengths from the satellites. The time for the signal to travel from the satellite to the earth is proportional to the path length, and since the velocity of the signal is known it is possible to calculate the path length. This information is computed by the processor in the receiver to give the position on the earth. If three satellites are being received, then it is possible to compute the latitude and longitude. If four satellites are being received, then it is also possible to compute the altitude as well.

The GPS satellites are controlled by a master station located at Falcon Air Force Base, Colorado Springs, USA. Other remote stations are located on Hawaii, Ascension Island, Diego Garcia and at Kwajalein. Using all

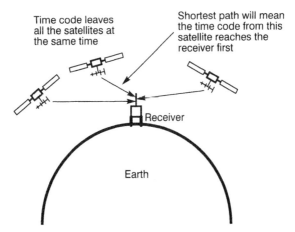

Time code leaves
all the satellites at
the same time

Shortest path will mean
the time code from this
satellite reaches the
receiver first

Receiver

Earth

Figure 8.9   *Operation of the Global Positioning System (GPS)*

these stations the satellites can be tracked and monitored for 92 per cent of the time. This results in two 1.5 hour periods each day when the satellite is out of contact with the ground stations. Using these stations the performance of the satellites is monitored very closely. The information that is received at the remote stations is passed to the main operational centre at Colorado Springs and the received information is assessed. Parameters such as the orbit and clock performance are monitored and actions taken to reposition the satellite if it is drifting even very slightly out of its orbit, or if any adjustment to the clock is necessary. This information is passed to three uplink stations co-located with the downlink monitoring stations at Ascension Island, Diego Garcia and Kwejalein

Weather and observation satellites

Satellites are used for a variety of other purposes. They can be used for a variety of observation tasks. One of the more familiar of these applications is in weather forecasting where they can take a bird's eye view of the weather formations from high above the earth. This gives weather forecasters a much better view of how the weather formations are developing and moving.

There are two main types of weather satellite. Geostationary ones are used for short-range warning. Polar orbiting satellites are used for longer-term forecasting. However, both types of satellite are needed to provide a complete picture.

The geostationary satellites are sufficiently high to gain a view of a large section of the earth's surface although their view of regions towards the poles is limited because of the low angle at which the satellites see these regions. As they remain over a certain point on the earth's surface they can provide a constant view of the prevailing conditions below them. They are usually used for monitoring trigger conditions for severe weather conditions such as hurricanes, tornadoes, thunderstorms and other forms of severe weather. The imaging equipment can also help to estimate the level of rainfall during a storm, as well as providing estimates of snowfall and providing estimates of spring snow melting.

The non-geostationary orbiting satellites are used for looking more closely at the earth's surface. Typically these satellites are in a polar orbit (i.e. crossing both poles) at an altitude of around 800–900 km. They contain instruments for monitoring the earth's atmosphere and the surface. They use both visible and infrared data to give radiation measurements, temperature profiles and the like. They also monitor incoming particles such as electrons, protons, etc., and using ultraviolet sensors they are able to provide indications of the ozone levels in the atmosphere and monitor the size of the ozone hole. Although not associated with weather some of these satellites perform another valuable task as they receive, process and then retransmit data from search and rescue beacons, so that people needing rescue can be located anywhere on the earth's surface. There have been several stories of miraculous rescues resulting from the use of these beacons.

9 Private mobile radio

Businesses of all types make widespread use of the radio spectrum, and have a great need for communications. Private mobile radio (PMR) has been in existence for many years, and differs from the more familiar mobile phone system in that access is restricted to those with licences. It is usually used for communication of a mobile station with its base, although in many instances two mobile stations can communicate. There are obvious examples like taxis that need to keep in contact with their bases to be able to pick up new customers. Other major users of this type of radio communications are the emergency services. However, there are many other users who make extensive use of this type of communication, and large sections of the VHF and UHF portions of the radio spectrum are allocated to these services.

Originally private mobile radio was used to cover a restricted area. A main base station was used to communicate with a number of mobile stations mounted in vehicles. This limited the coverage of the system to that of the base station. Systems were often fairly elementary by today's standards and transmissions used simplex mode, requiring press to talk on the transmitters. For many applications such as taxi services this was perfectly satisfactory as they would operate within a given service area and there was no need for a sophisticated system, especially as this would have added extra cost.

However, some operators did require much greater ranges and to accommodate this type of user, trunked PMR was developed. This allowed calls to be routed to transmitters in different areas, thereby significantly increasing the range. Systems have been considerably developed in recent years with the introduction of a system known as TETRA. Originally the letters stood for Trans European Trunked RAdio, but as the system is now being used beyond Europe the abbreviation now stands for TErrestrial Trunked RAdio.

Local private mobile radio

This is the simplest of the PMR systems and involves the use of a single antenna site for the location of the base station. In view of the fact that the antenna may be mounted on a high tower, coverage may extend up to distances of 50 kilometres, although ranges somewhat less than this are more usual.

Licences are allocated for operation on a particular channel or channels. The user can then have use of these channels to contact the mobile stations in his fleet. The base station may be run by the user himself or it may be run by an operating company who will hire out channels to individual users. In this way a single base station with a number of different channels can be run by one operator for a number of different users and this makes efficient use of the base station equipment. The base station site can also be located at a position that will give optimum radio coverage, and private lines can be provided to connect the users' control office to the transmitter site.

As there is no incremental cost for the transmissions that are made, individual calls are not charged, but instead there is a rental for overall use of the system. For those users with their own licences they naturally have to pay for the licence and the cost of the equipment.

While many systems operate with the remote or mobile stations being able to hear all the calls being made, this may not always be satisfactory and a system of selective calling may be required. There are two ways of achieving this. One is to use dual tone multiple frequency (DTMF) signalling whereas the other uses the continuous tone coded squelch system (CTCSS).

DTMF is a system that is widely used for telephone signalling and is almost universally used for touch tone dialling for landline telephones today. Set pairs of tones are used to carry the information. The eight frequencies used are 697, 770, 852 and 941 Hz (termed the 'low tones') and 1209, 1336, 1477 and 1633 Hz (termed the 'high tones'). One high and one low tone are used and the combinations are used to represent different numbers.

The relevant code consisting of one or more digits is sent and the station is programmed to respond to the number, typically one or two digits respond by opening the squelch on the receiver to let the audio through. The disadvantage of this system is that if the receiver does not pick up the code at the instant the DTMF signalling takes place then it will not respond to any of the message. This can be a significant disadvantage because mobile stations often lose the signal for short periods as they are on the move.

The other widely used system is CTCSS. This is also called sub-audible tones or PL tones (a Motorola trademark). It is a system where a subaudible tone (below about 250 Hz) is transmitted in addition to the normal voice channel. Only when the correct tone for the required station is transmitted will the squelch for that receiver be opened and the transmitted audio will be heard. The advantage of this system is that the sub-audible tones are transmitted for the whole period of the transmission so if the signal fades at the beginning of the transmission and then increases in strength, this part of the transmission will be heard. Systems typically are able to provide up to 37 different tones, the lowest frequency of which is 67 Hz and the highest 250.3 Hz.

In general narrow-band frequency modulation is the chosen form of modulation, although airport services use amplitude modulation. Typically a deviation of 2.5 kHz is used for FM and this enables a channel spacing of 12.5 kHz to be implemented. As the demands for PMR are high, it is necessary to make effective use of the channels available. This is achieved by reusing the frequencies in different areas. Base stations must be located sufficiently far apart so that interference is not experienced, and also selective calling techniques such as CTCSS and DTMF are used to ensure that as many mobiles as possible can use a given channel.

Trunked PMR

The elementary form of private mobile radio, although simple and relatively low cost to implement, has many limitations. It cannot provide for routing of incoming calls over a wide area. To implement a system of this nature adds a considerable degree of complexity and requires a network of base station sites. This in turn requires further complexity because the system that is set up needs to know the location of the mobile station so that any calls can be routed through to the relevant base station for onward transmission to the mobile. As a result of this, charging structures are different to those used for simple PMR, and are generally on the basis of utilization. Therefore each call is recorded for this purpose.

In view of the very high cost of setting up trunked networks, they are normally run by large leasing companies or consortia that provide a service to a large number of users. In view of the wider areas covered by these networks and the greater complexity, equipment has to be standardized so that suppliers can manufacture in higher volumes and thereby reduce costs to acceptable levels. Most trunked radio systems follow the same format. This is most widely known as MPT 1327.

To implement trunked PMR a network of stations is set up. These stations are linked generally using land lines, although optical fibres and point to point radio are also used. In this way the different base stations are able to communicate with each other.

In order to be able to carry the audio information and also run the variety of organizational tasks that are needed the system requires different types of channel to be available. These are the control channels of which there is one in each direction for each base station or trunking system controller (TSC). A number of different control channels are used so that adjacent base stations do not interfere with one another, and the mobile stations scan the different channels to locate the strongest control channel signal. In addition to this there are the traffic channels. The specification supports up to 1024 different traffic channels to be used. In this way a base station can support a large number of different mobile stations that are communicating at the same time. However, for small systems with only a few channels, the control channel may also act as a non-dedicated traffic channel.

The control channels use signalling at 1200 bits per second with fast frequency shift keying (FFSK) sub-carrier modulation. It is designed for use by two-frequency half duplex mobile radio units and a full duplex TSC.

For successful operation it is essential that the system knows where the mobiles are located so that calls can be routed through to them. This is achieved by base stations polling the mobile stations using the control channel.

To make an outgoing call the mobile transmits a request to the base station as requested in the control channel data stream from the base station. The mobile transmits its own code along with that of the destination of the call, either another mobile or a control office. The control software and circuitry within the base station and the central control processing area for the network sets up the network so that a channel is allocated for the audio (the traffic channel). It also sets up the switching in the network to route the call to the required destination.

To enable the mobile station to receive a call, it is paged via the incoming control channel data stream to indicate that there is an incoming call. Channels are allocated and switching set up to provide the correct routing for the call.

There is no method to 'handover' the mobile from one base station to the next if it moves out of range of the base station through which a call is being made. In this way the system is not a form of cellular telephone. It is therefore necessary for the mobile station to remain within the service area of the base station through which any calls are being made.

The control channel signalling structure has to be defined so that all mobiles know what to expect and what data is being sent. Signalling on the forward control channel is nominally continuous with each slot comprising 64-bit codewords. The first type is the control channel system codeword (CSCC). This identifies the system to the mobile radio units and also provides synchronization for the following address codeword.

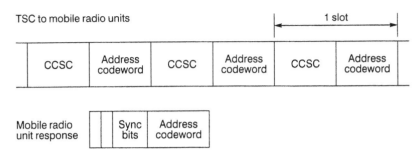

Figure 9.1 *The control channel signalling structures*

As mentioned the second type of word is the address codeword. It is the first codeword of any message and it defines the nature of the message. It is possible to send data over the control channel. When this occurs, both the CSCC and the address codewords are displaced with the data appended to the address codeword. The mobile radio unit data structure is somewhat simpler. It consists fundamentally of synchronism bits followed by the address codeword.

There are a number of different types of control channel messages that can be sent by the base station to the mobiles:

Aloha messages   Sent by the base station to invite and mobile stations to access the system.

Requests   Sent by radio units to request a call to be set up.

'Ahoy' messages   Sent by the base station to demand a response from a particular radio unit. This may be sent to request the radio unit to send his unique identifier to ensure it should be taking traffic through the base station.

Acknowledgements   These are sent by both the base stations and the mobile radio units to acknowledge the data sent.

Go to channel messages   These messages instruct a particular mobile radio unit to move to the allocated traffic channel.

Single address messages   These are sent only by the mobile radio units.

Short data messages   These may be sent by either the base station or the mobile radio unit.

Miscellaneous messages   Sent by the base station for control applications.

One of the problems encountered by mobile signalling systems is that of clashes when two or more mobile radio units try to transmit at the same time on the control channel. This factor is recognized by the system and is overcome by a random access protocol that is employed. This operates by the base station transmitting a synchronization message inviting the mobile radio units to send their random access message. The message from the base station contains a parameter that indicates the number of timeslots that are available for access. The mobile radio unit will randomly select a slot in which to transmit its request but if a message is already in progress then it will send its access message in the next available slot. If this is not successful then it will wait until the process is initiated again.

Although the data is transmitted as digital information, the audio or voice channels for the system are analogue, employing FM. However, some work has been carried out to develop completely digital systems. The main systems are by Motorola, by Ericsson (EDACS) and Johnson (LTR). These systems have not gained particularly widespread acceptance.

TETRA

Although trunked PMR provides many advantages over the simple PMR, advancing technology has enabled far more sophisticated systems to be envisaged. As a result the most recent PMR system called TETRA has been launched. Its specification is controlled by the European Telecommunications Standards Institute (ETSI), although it has gained use in many countries outside Europe.

TETRA provides a number of advantages over trunked private mobile radio. It uses digital speech transmission. This in itself provides a large degree of privacy against listeners, but in addition to this encryption is also supported, enabling it to be completely secure for services such as the police. Data transmissions can take place at a greater rate. The data rate for MPT 1327 trunked radio is 1.2 kbits per second whereas for TETRA it is significantly faster at 7.2 kbits per second for a single channel. This can be increased fourfold to 28.8 kbits per second when multislot operation is employed. It also enables efficient use of the available spectrum to be made with four slots per 25 kHz to be used under its time division multiple access (TDMA) system. This effectively means that each user occupies just 6.25 kHz bandwidth. The system also supports a number of other features including call hold, call barring, call diversion and ambience listening.

There are three different modes in which TETRA can be run. They are voice plus data (V + D), direct mode operation (DMO), and packet data optimized (PDO).

The most commonly used mode is V + D. This mode allows switching between speech and data transmissions, and can even carry both by using different slots in the same channel. Full duplex is supported with base station and mobile radio unit frequencies normally being offset by about 10 MHz to enable interference levels between the transmitter and receiver in the station to be reduced to an acceptable level.

DMO is used for direct communication between two mobile units and supports both voice and data; however, full duplex is not supported in this mode. Only simplex is used. This is particularly useful as it allows the mobile stations to communicate with each other even when they are outside the range of the base station.

Finally the PDO mode is optimized for data only transmissions. It has been devised with the idea that much higher volumes of data will be

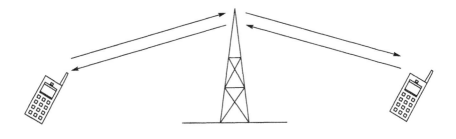

(a) V+D mode

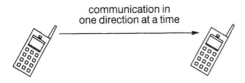

(b) Direct mode operation

Figure 9.2   *TETRA modes of operation*

needed in the future and it is anticipated that further developments will be built upon this standard.

TETRA uses TDMA enabling efficient use of the spectrum by allowing several users to share a single frequency. As the speech is digitized, both voice and data are transmitted digitally and multiplexed into the four slots on each channel. Digitization of the speech is accomplished using a system that enables the data to be transmitted at a rate of only 4.567 kbits per second. This low data rate can be achieved because the process that is used takes into account the fact that the waveform is human speech rather than any varying waveform. The digitization process also has the advantage that it renders the transmission secure from casual listeners. For greater levels of security that might be required by the police or other similar organizations it is possible to encrypt the data. This would be achieved by using an additional security or encryption module.

The data transmitted by the base station has to allow room for the control data. This is achieved by splitting what is termed a multiframe lasting 1.02 seconds into 18 frames and allowing the control data to be transmitted every 18th frame. Each frame is then split into four timeslots. A frame lasts 56.667 ms. Each timeslot then takes up 14.167 ms. Of the 14.167 ms only 14 ms are used. The remaining time is required for the transmitter to ramp up and down. The data structure has a length of 255 symbols or 510 modulation bits. It consists of a start sequence that is followed by 216 bits of scrambled data, a sequence of 52 bits of what is termed a training sequence. A further 216 bits of scrambled data follow and then the stream is completed by a stop sequence. The training sequence in the middle of the data is required to allow the receiver to adjust its equalizer for optimum reception of the whole message.

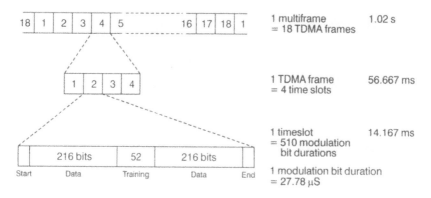

Figure 9.3 *TETRA data structure*

The data is modulated onto the carrier using differential quaternary phase shift keying. This modulation method shifts the phase of the RF carrier in steps of $\pm\pi/4$ or $\pm 3\pi/4$ depending upon the data to be transmitted. Once generated the RF signal is filtered to remove any sidebands that extend out beyond the allotted bandwidth. These are generated by the sharp transitions in the digital data. A form of filter with a root raised cosine response and a roll-off factor of 0.35 is used. Similarly the incoming signal is filtered in the same way to aid recovery of the data.

Beyond this TETRA uses error tolerant modulation and encoding formats. The data is prepared with redundant information that can be used to provide error detection and correction. The transmitter of each mobile station is only active during the timeslot that the system assigns it to use. As a result the data is transmitted in bursts. The fact that the transmitter is only active for part of the time has the advantage that the drain on the battery of the mobile station is not as great as if the transmitter were radiating a signal continuously. The base station, however, normally radiates continuously as it has many mobile stations to service.

One important feature of TETRA is that the call set-up time is short. It occurs in less than 300 ms and can be as little as 150 ms when operating in DMO. This is much shorter than the time it takes for a standard cellular telecommunications system to connect. This is very important for the emergency services where time delays can be very critical.

10 Cellular telecommunications

Today cellular telecommunications is an accepted part of everyday life. Its growth has been phenomenal and is seen as one of the technological success stories of the end of the twentieth century. This can be seen by the fact that in many countries some people have two or more phones. This growth has been seen across all areas of the population. Youngsters use their phones as much if not more than those people in other age brackets, indicating that in years to come use will remain very high as these people grow older. It is a well-known fact that the growth of cellular telecommunications has far outstripped analysts' expectations. Out of this many new companies have been created, and grown to rival some of the largest companies in the world in terms of size in just a very few years.

Evolution

The concept of cellular telecommunications dates back to 1947 at Bell Laboratories when an internal document proposed the idea of a system using a number of low power transmitters in 'cells' to enable frequency reuse in a mobile radio or telecommunications system. The proposal even mentioned the need for a method of 'handing over' the mobile station from one cell to the next as it moved along. However, the document does not state how this might be achieved. The system remained dormant for many years, and although a number of mobile phones were in use, these were relatively simple, effectively being a normal two-way radio system using one frequency for transmit and another for receive. On early models the user even had to use a press to talk switch. On top of this they only operated in populated areas, requiring a single transmitter to provide the required coverage. Naturally this was limited, as was the number of available channels. This caused problems as few people had access to them. In fact in the 1970s there were typically many times more people on the waiting list for these early mobile phones than were actually 'connected' and able to use them.

In the late 1960s and early 1970s a number of countries started to consider seriously the possibility of a cellular telecommunications system. In Japan, for example, the Nippon Telegraph and Telephone Company proposed a nationwide cellular system at 800 MHz. In Finland ideas also started to move forward. Then in the USA, the Bell Telephone Laboratories submitted a patent proposal in December 1970.

The next major development occurred in 1969 when a radio telephone system employing frequency reuse was used aboard a train. A total of six channels in several zones were used along the route that spanned over 200 miles with the system under computer control. However, it took until 1975 before the FCC gave approval for Bell to start a trial system, and two more years before it was allowed to operate. Not surprisingly the development of a new technology cost very significant amounts, many millions of dollars were spent, and eventually systems started to be seen. In fact the first development cellular telephone system began operation in May 1978 in Bahrain. Although relatively simple in some respects, the system had two cells and about 250 subscribers. However, development in the USA was moved ahead very swiftly and two months later in July 1978 the Advanced Mobile Phone Service (AMPS) commenced operation around Chicago. Initially the system was trialled using Bell employees, but in December of that year paying customers started to use the system. Then it took until 1983 before full commercialization of the system took place in the USA. However, the first mobile phone system to be launched commercially was the Nordic Mobile Telephone (NMT) which was launched in 1980 using a band of frequencies at 450 MHz.

Development in many parts of Europe followed on behind the USA. A system known as Total Access Communications System (TACS) developed by Motorola was used in many countries. In the UK licences were awarded in 1985. Two companies were given licences, one company was partly owned by the previously state owned British Telecommunications (BT), and the other was called Racal Vodaphone. Owned by Racal Electronics plc, this company was later floated off as a separate company to become Vodafone, now one of the world's largest mobile phone companies.

Naturally cellular telecommunications technology spread around the world to many countries, and several other standards were introduced. Although analogue systems worked well they had some drawbacks, and ideas for digital systems were forming. One of the first was a European initiative which started life as Groupe Spéciale Mobile. Its name was later changed to Global System for Mobile Communications, and the initials GSM were retained. Initial work for this started in 1982. A total of 26 telecommunications companies within Europe co-operated on the development of the new system and it commenced operation with frequencies in the 900 MHz band in mid-1991. The same basic system is also used at

1800 MHz where it is often known as DCS 1800 (Digital Communications System) or GSM 1800, and in North America at 1900 MHz where it is called PCS 1900 (Personal Communications System) or GSM 1900. It is also being used increasingly in an 850 MHz band in North America.

With GSM established the next major development it underwent was the introduction of the short message service (SMS). Initially thought to be an interesting development, its use rose rapidly, especially as young people found it a cheap way of communicating using their phones.

Meanwhile in 1994 in the USA a company named Qualcomm proposed a system based around a spread spectrum technique, using CDMA. The system was given the trade name cdmaOne and its specification was classified as IS-95. The system possessed several advantages and gave an increase in capacity.

As the 1990s came to a close the cellular phone industry was booming. Industry analysts reasoned that people would want to use far more data services as they saw a significant rise in the use of the Internet. Existing systems were not able to support sufficiently fast data services and new systems were sought. The first step on the way was known as the General Packet Radio System (GPRS) and its enhanced system Enhanced Data rates for Global Evolution (EDGE). These systems were dubbed 2.5G as they were a development of the second generation system. However, the main goal was a fully third generation system. Three 3G systems have emerged. In Europe a system known as the Universal Mobile Telecommunications System (UMTS) using wideband CDMA (W-CDMA) appeared. In the USA a system known as CDMA2000 was adopted. This provided an evolutionary path from cdmaOne through to the full 3G standard with backward compatibility. The first commercial launch of CDMA2000 was in October 2000 in South Korea with the CDMA2000 1x system. A third system known as Time Division Synchronous CDMA (TD-SCDMA) was developed in China.

Basic standards

Looking at how the mobile telecommunications market has developed over the years it can be seen that a large number of standards have arisen, all with their own abbreviations. It is useful to summarize these to understand their salient features and where they are used. As already mentioned there are the three generations of system, generally known as 1G, 2G and 3G to denote the different generations. There is also 2.5G that is a development of the second generation phones, but not fully to the 3G standards. Another important feature of the systems is the way in which different users are given access to the system. In some different frequencies are used. This is called frequency division multiple access

Table 10.1

System	Generation	Channel spacing	Access	Comments
AMPS	1G	30 kHz	FDMA	Advanced Mobile Phone System, this analogue system first developed and used in the USA
NAMPS	1G	10 kHz	FDMA	Narrow-band version of AMPS chiefly used in the USA and Israel based on a 10 kHz channel spacing.
TACS	1G	25 kHz	FDMA	Analogue system originally in the UK. Based around 900 MHz, this system spread worldwide. After the system was first introduced, further channels were allocated to reduce congestion, in a standard known as Extended TACS or ETACS
NMT	1G	12.5 kHz	FDMA	Nordic Mobile Telephone. This analogue system was the first system to be widely used commercially. Used initially on 450 MHz and later at 900 MHz. It was used chiefly in Scandinavia but it was adopted by up to 30 other countries.
GSM	2G	200 kHz	TDMA	Originally called Groupe Spéciale Mobile, the initials later stood for Global System for Mobile communications. It was developed in Europe and first introduced in 1991. The service is normally based around 900 MHz although some 850 MHz allocations exist in the USA.
DCS 1800	2G	200 kHz	TDMA	1800 MHz derivation of GSM and is also known as GSM 1800.

System	Generation	Bandwidth	Access	Description
PCS 1900	2G	200 kHz	TDMA	1900 MHz derivation of GSM and is also known as GSM 1900.
USDC	2G	30 kHz	TDMA	US Digital Cellular. This system was introduced in 1991 and is sometimes called North America Digital Cellular – it is also known by its standard number IS-54 that was later updated to standard IS136. As it is based on a TDMA system it is often just called 'TDMA' in the countries where it is used. It is a 2G digital system that was designed to operate alongside the AMPS system.
GPRS	2.5G	200 kHz	TDMA	General Packet Radio Service. A data service that can be layered onto GSM. It uses packet switching instead of circuit switching to provide the required performance. Data rates of up to 115 kbps are attainable.
EDGE	2.5/3G	200 kHz	TDMA	Enhanced Data rates for Global Evolution. The system uses a different form of modulation (8PSK) and packet switching which is overlaid on top of GSM to provide the enhanced performance. Systems using the EDGE system may also be known as EGPRS systems.
cdmaOne	2G	1.25 MHz	CDMA	This is the brand name for the system with the standard reference IS95. It was the first CDMA system to gain widespread use. The initial specification for the system was IS95A, but its performance was later upgraded under IS95B. Apart from voice it also carries data at rates up to 14.4 kbps for IS95A and under IS95B data rates of up to 115 kbps are supported.

Table 10.1 *continued*

System	Generation	Channel spacing	Access	Comments
CDMA2000 1X	2.5G	1.25 MHz	CDMA	This system supports both voice and data capabilities within a standard 1.25 MHz CDMA channel. CDMA2000 builds on cdmaOne to provide an evolution path to 3G. The system doubles the voice capacity of cdmaOne systems and also supports high speed data services. Peak data rates of 153 kbps are currently achievable with figures of 307 kbps quoted for the future, and 614 kbps when two channels are used.
CDMA2000 1xEV-DO	3G	1.25 MHz	CDMA	The EV-DO stands for Evolution Data Only. This is an evolution of CDMA 2000 that is designed for data only use and its specification is IS 856. It provides peak data rate capability of over 2.45 Mbps on the forward or downlink, i.e. from the base station to the user. The aim of the system is to deliver a low cost per megabyte capability along with an always on connection costed on the data downloaded rather than connection time.
CDMA2000 1xEV-DV	3G	1.25 MHz	CDMA	This stands for Evolution Data and Voice. It is an evolution of CDMA2000 that can simultaneously transmit voice and data. The peak data rate is 3.1 Mbps on the forward link. The reverse link is very similar to CDMA2000 1X and is limited to 384 kbps.

UMTS	3G	5 MHz	CDMA/TDMA	Universal Mobile Telecommunications System. Uses Wideband CDMA (W-CDMA) with one 5 MHz channel for both voice and data, providing data speeds up to 2 Mbps.
TD-SCDMA	3G	1.6 MHz	CDMA	Time Division Synchronous CDMA. A system developed in China to establish their position on the cellular telecommunications arena. It uses the same bands for transmit and receive, allowing different timeslots for base stations and mobiles to communicate. Unlike other 3G systems it uses a time division duplex (TDD) system.

(FDMA). Another uses different timeslots or time division multiple access (TDMA). A third multiple access scheme used code division multiple access (CDMA) where spread spectrum techniques are used. Here different orthogonal spreading codes are used to distinguish between the different users and allow them access to the system.

Basic concepts

As the name implies the basic concept around which cellular tele-communications operates is that the coverage area is split up into small areas called cells. Each one is covered by a relatively low power base station. This has advantages over the earliest concepts used by relatively high power mobile phones where a single high power station was used to provide coverage over a large area. The cell system enables local area coverage using low power handsets so that at a certain distance away the frequencies can be reused. Interference is kept to a minimum by the fact that adjacent cells use different frequencies or channels, although a different technique is used for a CDMA system as we shall see later. By arranging that the mobile handset is able to change from one of the channels supported by one cell to one of the channels supported by the adjacent one, virtually seamless coverage can be achieved. While this creates additional complexity in the base station and handset, it enables much greater use to be made of the available channels.

Different arrangements of cells may be used dependent upon the requirements. These may vary according to the terrain and the level of usage. The cells are arranged in groups that are termed clusters. The most typical is a seven cell cluster like that shown in Figure 10.1. This is very much an idealized view because coverage never stops exactly at the edge of the cell and the shapes vary according to the terrain. Accordingly other formats including four, 12 and even 21 cell clusters may be found.

As there are only a limited number of channels available, the balance between the number of channels used by each base station and the number of cells in a cluster has to be made. If few cells are used in each cluster, then the channels have to be reused more frequently, i.e. fewer cells in a cluster mean that the distances between the two base stations that use the same channels are smaller and interference may be a problem. However, it has the advantage that greater numbers of channels are available in each cell, and hence there is the capacity to handle a greater number of calls. However, if the number of cells in a cluster is increased to reduce the level of interference, the larger number of cells has to share the same number of channels and each base station has fewer channels and the capacity is reduced.

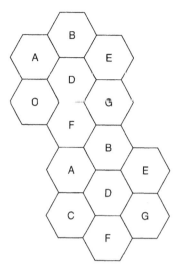

Figure 10.1 *An idealized cell system*

The different types of cells are given different names. Macrocells are large cells that are usually used for remote or sparsely populated areas. These may be 10 km or possibly more in diameter. Microcells are those that are normally found in densely populated areas which may have a diameter of around 1 km. Picocells may also be used for covering very small areas such as particular areas of buildings, or possibly tunnels where coverage from a larger cell is not possible. Obviously for the small cells, the power levels used by the base stations are much lower and the antennas are not positioned to cover wide areas. In this way the coverage is minimized and the interference to adjacent cells is reduced.

Other types of cell may be used for some specialist applications. Sometimes cells termed selective cells may be used where full 360 degree coverage is not required. They may be used to fill in a hole in the coverage, or to address a problem such as the entrance to a tunnel, etc. Another type of cell known as an umbrella cell is sometimes used in instances such as those where a heavily used road crosses an area where there are microcells. Under normal circumstances this would result in a large number of handovers as people driving along the road would quickly cross the microcells. An umbrella cell would take in the coverage of the microcells (but use different channels to those allocated to the microcells). However, it would enable those people moving along the road to be handled by the umbrella cell and experience fewer handovers than if they had to pass from one microcell to the next.

Transmissions both ways

Basic radio communications systems use a single channel and what is known as a press to talk system, where the user presses a button or 'pressel' on the microphone to talk. This system is known as simplex as it uses a single channel. For a phone system a full duplex system is required where it is possible to speak in both directions at the same time. There are two main ways in which this can be achieved. The first is to transmit in one direction on one frequency and simultaneously transmit in the other direction on another. To achieve this there must be sufficient frequency separation and very good filters to make sure that the transmitter does not interfere with the receiver. This system is known as Frequency Division Duplex (FDD).

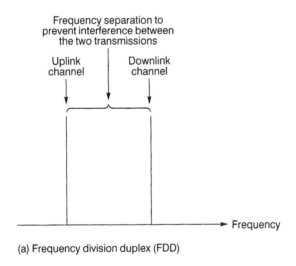

(a) Frequency division duplex (FDD)

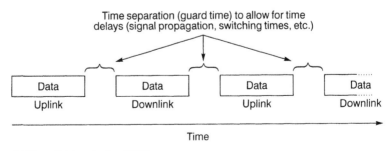

(b) Time division duplex (TDD)

Figure 10.2 *Representation of frequency and time division duplex*

The other system uses only a single frequency and can be employed where digital or data systems are used. This requires the analogue audio signal to be digitized. A single frequency is used for the radio frequency signal and short packets of data are sent first in one direction and then the other. As these data bursts are relatively short the user does not notice the short delay introduced by the fact that the digitized speech signal is not sent immediately. This type of system is known as Time Division Duplex (TDD).

It is often necessary to distinguish between the link from the mobile to the base station and the link from the base station to the mobile. The first, i.e. the link from the mobile to the base station, is often called the uplink or the reverse link as the signal is being transmitted up to the base station. The second, i.e. the link from the base station to the mobile, is known as the downlink or the forward link.

Mobile phones

Today's mobile phones or handsets are remarkable feats of electronic engineering. They contain a vast amount of functionality in a very small space. In fact the requirements for more functionality in smaller packages has forced the pace of technology development, particularly in the radio frequency arena. Phones contain a receiver, transmitter and all the logic and control circuitry (which will undoubtedly be based around a

Figure 10.3 *A typical mobile phone handset (courtesy Sony Ericsson)*

microprocessor) required to maintain the call as well as an advanced user interface consisting of a keypad and display. To enable these to perform all the functions that are required and to send and receive all the correct codes and controls, a vast amount of software is required. Without this the phone would not be able to operate.

There are a number of different topologies that can be adopted for phones. The first generation phones typically used a double conversion superhet, converting down from the incoming signal frequency to a first IF that was usually around 45 MHz and then to a second IF at either 10.7 MHz or 455 kHz. However, today's phones generally convert right down to the baseband frequency where the demodulation takes place. Obviously the transmit side is the inverse of the receiver. This approach significantly simplifies the design, and hence there is less circuitry.

The local oscillator signal is generated using a frequency synthesizer. In this way the internal processing and logic in the phone can accurately and easily control the transmitter and receiver frequencies. This is required so that the base station can control the channels to be used.

Batteries for phones are becoming increasingly important. All use packs of rechargeable cells. Very old ones might still use nickel cadmium (NiCad) packs, but many of the new handsets use nickel metal hydride (NiMH) or lithium ion (Li-ion) ones. The new cells can store more charge in a given volume or weight and can therefore last longer. However, they do need to be charged correctly as they are more easily damaged than their NiCad counterparts. New battery developments mean that new types of battery are appearing. The longer battery life is becoming increasingly important because the higher data content can result in increased battery drain and reduced battery life between charges. Comparing the earliest phones with those used today, enormous changes can be seen. Some of the first cellular phones required very large batteries by today's standards. These were required for three reasons. The first was that the integrated circuits used required relatively high levels of current compared to those available these days. Second, the large cell sizes of these early networks dictated that high power levels were needed, and this reflected in greater levels of current consumption. Finally, as already mentioned, battery technology has improved considerably.

Base station

The base station consists of a number of different elements. The first is the transceiver itself. This contains the electronics for communicating with the mobile handsets. Second, there is the antenna and the feeder to connect the antenna to the base station itself. These antennas are visible on top of masts and tall buildings enabling them to cover the required area. Finally, there is the interface between the base station and its controller further up the network. This consists of control logic and software as well as the cable link to the controller.

The base stations are set up in a variety of places. In towns and cities the characteristic antennas are often seen on the top of buildings, whereas in the country separate masts are used. It is important that the location, height and orientation are all correct to ensure the required coverage is achieved. If the antenna is too low or in a poor location, there will be insufficient coverage and there will be a coverage 'hole'. Conversely if the antenna is too high and directed incorrectly, then the signal will be heard well beyond the boundaries of the cell. This may result in interference with another cell using the same frequencies.

Figure 10.4 *A typical cellular telecommunications base station along the roadside*

The performance of the equipment in the base station must make up for many of the deficiencies in the mobile set. The base station will often have an output power of around 10 watts. The receiver must be sensitive, the site must be free from high levels of noise which might mask out the wanted signals, and the antennas must have gain to enable the cell signal to be heard throughout the cell, and to enable it to receive signals in the cell. The mobile or portable set on the other hand has a limited power, the antenna will be small and not offer a good performance, and the location may be very poor as mobile phones are often used inside buildings or in other poor radio locations.

Infrastructure

Although the phones and base stations are the obvious elements of a mobile telecommunications system, there is a considerable amount of infrastructure behind the base stations that are equally as important. Its operation enables the whole system to come together and operate seamlessly giving the user a high quality of performance. While the

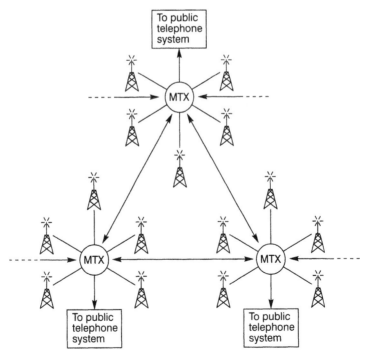

Figure 10.5 *A simplified version of a cellular phone system*

different phone standards employ different types of infrastructure many of the same basic concepts are applicable to all types of network.

For the cellular system to operate satisfactorily the base stations must be able to route calls through to the ordinary telephone network as well as contacting other mobile phones. Additionally they must be able to control the phones within the cell and pass them on to the next cell when the phone moves out of their service area into the next. All this requires a high degree of control.

To enable all the switching and control functions to take place the base station is linked to control and switching centres. To avoid all this being routed back to very large centralized centres, some of this is done locally. Typically there are centres sometimes called mobile telephone exchanges (MTX) or mobile switching centres (MSC) that control a number of base stations and smaller localized controllers often called base station controllers (BSC) that might control a small group of cells, possibly a complete cluster. Monitoring and routing of the mobile phone is first accomplished by the local base station controller. Once the mobile is about to move out of the set of base stations controlled by the localized controller it hands control to the larger switching centre. However, it should be remembered that different mobile phone systems may operate in slightly different ways although the underlying principles are the same.

Control

In order that a cellular phone system is able to work, the base station has to be able to control many aspects of the operation of the phone. Which frequencies are to be used, knowledge of where the mobile handset is so that calls can be routed to it, and many other functions all need to be accomplished for it to be able to operate correctly. In order to achieve these control functions most systems have dedicated control channels that are used purely for this purpose and they do not carry any voice or other user data.

The first function that must be completed when the mobile handset is turned on is that of registration. After the phone has performed its own internal power on self-check it scans the channels for the strongest control channel. Once it finds this it identifies itself on the uplink to the base station. Data including the individual phone ID and the number is sent regardless of whether a call is to be made. This information is then recorded back at the main system. In this way the system can register that the phone is within a particular cell so that it can be located for incoming calls. The phone then identifies itself periodically after this in case it moves from one cell to another. It also repeats this action if it detects it has moved from one cell to another.

The other function that has to be undertaken is that of handover. If the mobile handset moves from one cell to another it needs to be able to undertake a number of actions so that the handset communicates with a second base station instead of the first. This handover needs to be implemented very quickly so that there is no interruption to the communication otherwise the users will notice a disruption to their conversation. This handover is co-ordinated by the base station controllers and the mobile switching centres. For handover within a cluster local switching centres can be used to control this but when transfer takes place between clusters it is handled by switching centres further up the chain.

The system monitors the signal strength of the mobile. When it is falling in strength and the mobile signal can be heard better in the adjacent cell, the system arranges for it to swap to a clear channel supported by the new cell. When the change takes place the mobile swaps its frequency and the data from the base station is switched from the old base station to the new one and to the correct channel. All of this takes a fraction of a second and normally goes totally unnoticed by the user, especially when digital systems are used.

Analogue systems

The first mobile phone systems to be introduced were analogue. They were introduced at a time when IC technology was not as far advanced as it is today, and even then the first phones to be introduced were very large by comparison to those used today. In view of this the modulation schemes were relatively simple using FM for the voice channel and a form of frequency shift keying for the data channel. When compared to the systems in use today the spectrum usage was not particularly efficient. They also had the disadvantage that it was very easy for people with scanners to listen in to conversations, a fact that several people, including some who are very famous, discovered to their cost.

Analogue systems require voice and control channels in each direction for their operation as they operate full duplex. This gives a total of four channels as shown in Figure 10.6. The voice channels use narrow-band FM as this gives good immunity to general noise and signal strength variations. The capture effect exhibited when receiving FM signals also gives some immunity from signals from distant cells, thereby reducing the possibility of intercell interference. The data channels use frequency shift keying to carry the information to monitor the status of the phone while it is idle (switched on but not in use), to initiate a call and to allocate the voice channel.

Problems may be encountered with the signalling or data functions. Here multipath propagation, where the signal may reach the receiver via

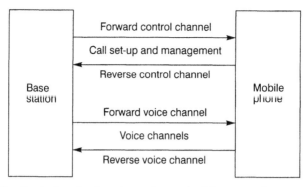

Figure 10.6 *Control and voice channels required for an analogue system*

different paths and at slightly different times, means that errors are often present. Corruption becomes worse when signal strengths are low. To overcome this various forms of error detection and correction are employed.

In order that many subscribers can be serviced within each cell there are a number of channels that can be used. These are split into two types: those for transmitting to the base station from the mobile phone, and those transmitting from the base station to the mobile station. These are split into two bands and arranged so that there is a sufficient separation between the transmitter and receiver signal frequencies. This is required because the same mobile handsets and base stations have to handle signals in both directions at the same time. It is obviously very important that interference should not be caused by the transmitted signal to the received one. Particularly the receiver must not be overloaded by the nearby strong transmitted signal.

GSM

Although the GSM system has many similarities with the basic concept for a cellular telecommunications system, there are naturally many improvements that have been made. To enable different organizations to be able to develop compatible pieces of equipment, the GSM specifications were set down. The original set of documentation extended to over 8000 pages, which although lengthy enabled the development to proceed relatively smoothly and to be adopted by many countries and organizations.

The system supports several features that were not available on the old analogue services. The one that has gained the most popularity is the short message service (SMS) that is very widely used, especially by the young for whom it is a cheap way of communicating.

The mobile station (MS) consists of the mobile radio transceiver with its display and the digital processing circuitry. It also contains a Subscriber Identity Module (SIM). This is a small card contained in the phone that enables the personal information set up on the phone to be transferred to another phone. It is particularly useful when upgrading phones. By transferring the card from one phone to another it enables items such as phone number, phone book and the like to be transferred.

Two identities are associated with the phone. One is the International Mobile Equipment Identity (IMEI) and this is contained in the phone itself. The other is the International Mobile Subscriber Identity (IMSI) and this is contained in the SIM card. The IMEI and IMSI are independent and this allows for both the phone and the user to be identified separately.

The base station subsystem (BSS) comprises two sections. One is the base station transceiver sub-system (BTS) and the other is the base station controller (BSC). Often one controller is used to control a few transceivers. They communicate over a specified interface known as the A_{bis} interface. By having a standard specification different suppliers can be used for each.

One link further up the chain from the base station controller is the Mobile Switching Centre (MSC). This is defined in the GSM specifications

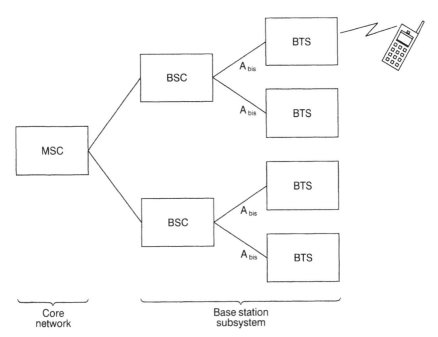

Figure 10.7 *Basic base station configuration for a GSM system*

Figure 10.8 *A typical cellular telecommunications base station located on a tall building covering an area of a town*

to act like a normal ISDN, or PSTN switching node. In addition to this it provides all the functionality to provide subscriber registration, authentication, call routing and handovers. The MSC also has access to the Authentication Centre, a protected database that stores a copy of the secret key kept in each subscriber's SIM card. It also checks the IMEI to ensure that the phone has not been reported stolen and that it is type approved. Only when the checks have been accomplished is the phone allowed onto the network.

The Home Location Register (HLR) is an important register. It is a large database in the system that stores the administrative information for each subscriber in a particular network. It details the location, i.e. the base station with which the mobile will communicate, whether the mobile is turned on and the services to which the user has access. To enable roaming, which is a key element of GSM, a further register called the Visitor Location Register (VLR) is provided to enable visitors to register on a network, for example when they are abroad. At registration the VLR obtains information from the user's HLR so that while the user is within the visited network he can be treated like a local user.

The GSM standard also defines the frequencies to be used: 890–915 MHz is used for the uplink and 935–960 MHz is used for the downlink. In some countries the full allocation is not available because other users require some of the frequencies in the band. Other similar systems, namely PCS 1800 and DCS 1900, use bands around 1800 MHz and 1900 MHz but adopt the same basic formats as GSM in other respects.

In order to make the optimum use of the channels available, GSM uses a combination of time and frequency division multiple access techniques. To enable the multiple frequencies to be used, the 25 MHz bandwidth is split into 124 carriers each of 200 kHz bandwidth. One or more of these can then be allocated to each base station, dependent upon the expected traffic. Then further capacity is gained using time division techniques. Each carrier is split into eight 'bursts' each of which lasts approximately 0.58 ms and each one is allocated to a user. The group of eight bursts is called a frame and this lasts 4.62 ms after which the sequence repeats, allowing an almost constant flow of data for each user.

As GSM is a digital system the speech has to be sampled and digitized. If this were linearly digitized it would require a data rate of around 64 kbps which is far too high as it would occupy too much of the available bandwidth. Accordingly the system that is used with GSM uses a form of prediction to help reduce the data rate. The actual system used is known as Regular Pulse Excited Linear Predictive Coder, and this gives a good balance between speech quality, data rate reduction, bandwidth usage, and also the complexity of processing required as this has an impact on battery usage. More complicated systems require more processing which in turn requires more circuitry and more current. In the system, the speech is sampled and encoded as 260 bits and this gives a total data rate of 13 kbps.

The data stream has error redundancy added to reduce bit errors. This data is then modulated onto the carrier using Gaussian filtered Minimum Shift Keying (GMSK). This was used because it gave a good compromise between transmitter complexity, efficient use of the available spectrum and good performance in terms of spurious emissions.

Frequency hopping is incorporated into the GSM specification. When this facility is used, the mobile handset and the base station change frequencies or hop from one frequency to another. This is accomplished between transmitting and receiving the sequential TDMA time frames. In other words the mobile handset will receive on one frequency, say frequency A, in the receive band, transmit on frequency R in the transmit band, then receive on frequency B in the receive band, and so forth. It can do this as it does not transmit and receive at the same time. The hopping algorithm allowing the mobile handset to know where to transmit and receive is broadcast on the Broadcast Control Channel. Frequency hopping has a number of advantages. Any interference on a given channel is present only for a short time, making the interference less noticeable, and any multipath effects are also reduced as multipath fading is dependent upon the frequency.

Another feature of GSM is called discontinuous transmission (DTX). It is found that during a conversation a person speaks for less than 40 per cent of the time. This means that usage of the uplink and downlink channels is poor. To improve this, and to reduce the level of general interference, the transmitter can be turned off when no speech or other useful sound is being carried. The voice activity detection has to detect the difference between extraneous noise and speech and correctly interpret the difference otherwise the speech will sound very clipped. Alternatively if it does not switch off often enough then improvement gained by using DTX is considerably reduced. A further addition is that during the dead periods no sound at all would normally be heard and this would sound very unusual. To overcome this and give the user confidence that the connection was not dead, some 'comfort' noise matching the background noise is added at the receiving end. The DTX solution that is used today took a considerable amount of work during its development and it can be very effective.

The power used by the system is tightly controlled. There are five classes of mobile station, and these are related to their output power. The highest is rated at 20 watts, but others are 8, 5, 2 and 0.8 watts. To minimize the levels of interference and also to conserve battery power both base stations and mobiles are able to step the power down to 20 milliwatts (13 dBm) so that the lowest power that can maintain acceptable communications can be used. To assess the required levels the mobile checks the signal quality and passes the information to the base station, which also assesses its received transmission quality. The base station alters its level and sends control data to the mobile so that it can set its output level accordingly.

The short message service (SMS) is particularly useful. Allowing text only messages of up to 160 characters it provides additional flexibility to the service and its unprecedented growth and popularity can be judged by the fact that in the year 2000 over six billion messages were sent every month. Although originally associated with GSM, other networks including CDMA and TDMA now support the service meaning that it is an almost universal mobile data service.

At the centre of the SMS system is the Short Message Centre (SMC). This is the entity that stores and forwards the messages for the networks. Each network has a gateway mobile switching centre (GMSC), and this is the mobile network's point of contact with the other networks. Messages are passed to and from the SMC from each of the network's GMSCs. When a message is received by the SMC it determines to which network it should be routed, and it enters the gateway MSC for that network. From here the network detects its location in the normal way and routes the message accordingly. Data for the SMS messages is passed over the signalling channels using a header to identify it as a short message. In this way messages can be received even when a voice call is in progress.

The short message service is being enhanced by the introduction of the Enhanced Message Service (EMS). This enables formatted text, pictures and sounds (melodies) to be sent. It uses the same basic format as SMS but larger data elements are sent and the header format is slightly changed. Further improvements in the service include a multimedia message service (MMS) which allows full colour pictures, animations and the like to be sent.

GPRS and EDGE

GPRS or General Packet Radio Service is a packet-based service that enables the GSM system to carry data at rates up to 114 kbps although normal data rates of around 30 to 50 kbps are likely to be experienced. In addition to this it will give a continuous connection to the Internet. It uses the same modulation (GMSK) as GSM and complements the existing services such as SMS. To upgrade the system to carry GPRS it is necessary to convert the system from what is termed a circuit switched network where the circuit is switched for the period when the user requires a particular circuit, to a packet switched network. The handsets, however, cannot be converted and new handsets are required when upgrading from GSM to GPRS.

Packet switching is the key to GPRS and it enables the system to be used only when data is being sent or received. Instead of devoting a complete channel to a mobile user for the duration of the call, as is normally the case for GSM, the channel is only used when data is actually being sent and as a result the available radio resource is shared between several users. Overall this means that large numbers of GPRS users can potentially share the same bandwidth and be served from a single cell. The actual number of users that can be supported at any time naturally depends upon how much data is being transferred.

A further update that can be used with GSM is known as EDGE. This stands for Enhanced Data rates for Global Evolution, and it is a standard originally called GSM 384. Often systems using EDGE are known as Enhanced GPRS or EGPRS systems. The system enables data transmission speeds of up to 384 kbps to be attained. However, to achieve this it must use all eight timeslots that can each carry 48 kbps. Additionally it uses a different form of modulation to that used for GSM. Instead of using GMSK it uses 8 phase shift keying modulation (8PSK) and this enables operators to offer similar data speeds to those that will be achieved with the full 3G networks. As such it provides an evolutionary path between the second and third generation networks. Implementation requires software and hardware upgrades to the base stations themselves and to the base station controllers. New handsets are also required.

cdmaOne

cdmaOne is the marketing name for the system that is specified under IS95. After it was first issued the specification was updated from IS-95A to IS-95B. The original specification supported voice and data up to 14.4 kbps; the updated IS-95B specification supports data up to 115.2 kbps.

Much of the way in which the system operates, and the infrastructure that is uses, is similar to GSM and will not be described again. However, the main difference is that it uses direct sequence spread spectrum in the form of CDMA as its modulation scheme. The use of different orthogonal codes (Walsh codes) provides the method of multiple access whereas GSM primarily uses TDMA and analogue systems use FDMA. It is claimed that CDMA systems provide a far more efficient use of the available spectrum than either FDMA or TDMA systems and for this reason CDMA is being widely adopted.

When using CDMA, all users in the cell operate on the same centre frequency. As the different users are allocated different orthogonal codes, each mobile is able to isolate its own signal from the base station and extract the data contained within the transmission. A full description of the way in which CDMA works is given in Chapter 3 and the information will not be repeated here. However, for the system to be able to operate satisfactorily the different signals being received by the base station have to be at approximately the same level. For this reason power level information is sent by the base station to each individual mobile to enable it to set its power output satisfactorily.

A further difference between CDMA systems and FDMA or TDMA systems is that base stations can operate on the same frequency. In other words the system can have a cluster size of 1. It is recognized that transmissions from neighbouring cells will cause interference and this will reduce the capacity of the system somewhat. Nevertheless there is still an improvement in performance.

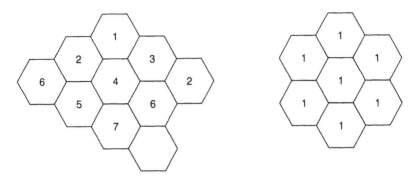

(a) Frequency reuse with FDMA system   (b) Frequency reuse with CDMA system

Figure 10.9 *Cellular system using CDMA to provide a cluster size of 1*

As the number of users on a given channel increases, the level of interference naturally increases and as a result there is a limit to the number of users that can be successfully supported on a given channel. Accordingly using different channels on different frequencies provides further capacity. In the case of cdmaOne, the channels are spaced 1.25 MHz apart, and in this sense there is some element of frequency division multiplexing.

To understand the operation of the system it is necessary to look at the transmissions that are made by the system. First, look at the downlink transmission, i.e. the transmission from the base station to the mobile. This consists of three common channels and the traffic channels. These are created by using different orthogonal codes within the main overall channel to or from the base station.

The three common channels consist of the pilot channel used to give the mobile an estimate of the path loss between it and the base station. In this way it can set its power level accordingly. It also carries synchronization information, as well as frequency offset information to enable the mobile to correct the receiver frequency.

The second control channel is the synchronization channel. This is primarily used by the mobile to acquire the timing reference. The timing is very important as the base stations are kept in tight synchronism with each other by a signal derived from the Global Positioning System (GPS). Each base station has a fixed timing offset from the GPS signal to minimize interference between cells and the synchronization channel provides the mobile with the required timing offset information to access the cell.

The third channel is the paging channel. This provides system parameters, voice pages, short message service and any other broadcast messages to users in the cell.

The traffic (radio or dedicated) channels carry the digitized speech and data. There are two types of traffic channel, namely the fundamental channel and the supplemental channel. Speech data is always transmitted over the fundamental channel and then between one and seven supplemental ones can be used to provide the additional capacity needed for data transmissions. Power control information from the base station to the mobile is also transmitted over the traffic channel. It enables the mobile to accurately control its power so that the signal received by the base station is within its required limits. If it becomes too strong, it will mask out the other mobile stations, preventing their signals from being decoded. If it is too weak, the base station will not be able to decode the data.

The speech is encoded into a digital format using a similar encoder to that used for GSM, and once encoded, error correction information is applied. This brings the modulation symbol rate to 19.2 kbps. This is then multiplied by the orthogonal code, known as a Walsh code, to give a 1.228 Mbps signal that is then transmitted. The transmitted signal occupies a channel of 1.25 MHz.

The uplink from the mobile to the base station is constructed slightly differently. Speech is generated in the same way, but more error correction is applied and this brings the modulation symbol rate up to 28.8 kbps. The signal is then spread using a Walsh code to bring the intermediate data rate up to 307 kbps. This does not provide sufficient spreading and it would not be able to counteract the levels of interference encountered. As a result the signal is further spread using another sequence known as a PN code. This multiplication further spreads the intermediate data by a factor of 4, resulting in a data rate of 1.2288 Mbps, the same as the rate transmitted by the base station.

The reason that the uplink and downlink are treated differently is that Walsh codes need to be synchronized if they are to remain orthogonal. As the signals transmitted from the mobile stations travel over different distances because of the variety of locations of the mobiles they will all arrive at slightly different times, and hence they will not be synchronized to one another. The addition of the PN code overcomes this problem, even though Walsh codes have been used for part of the spreading process in the mobile.

One of the advantages of CDMA systems is associated with handover. As a result of the fact that the same frequency is used for adjacent cells, mobiles are able to receive signals from more than one base station at the same time. Normally they would reject all but the signal from the wanted base station as part of the decoding process. However, the ability to receive more than one signal is advantageous when handing over from one cell to the next. During the handover process the two adjacent cells both send the same data signal to the mobile. During the handover process the mobile listens to both signals, and when the handover is complete the mobile returns to listening to just the one base station. This approach significantly reduces the possibility of a mobile being lost or dropping reception during handover, and it also ensures optimum voice quality when a mobile is on the edge of a cell. This form of handover is known as a 'soft' handover or handoff.

While soft handover appears to be a significant advantage there is a downside to the process. The first is that it reduces the system capacity. The reason for this is simply that it requires two channels or codes, one for each base station. It is estimated that this can reduce system capacity by as much as 40 per cent, but this is dependent upon many factors including the duration of the handover, the size of the cells, and the level of overlap between the cells. There are disadvantages for the mobile as well. It requires two receivers or demodulators to be able to monitor the two cells. While this adds complexity to the mobile, it can be overcome by integrating the additional functions into the main IC, and therefore the cost is relatively small. Despite these disadvantages, the advantages of a much smoother and more reliable handover are worth the additional costs.

Although one of the main advantages claimed for CDMA is a vast improvement in the efficiency of the use of the spectrum, this parameter is very difficult to calculate in real terms. This is because it depends to a large extent on a variety of aspects including the level of interference from outside the cell, the cell density and such like. Even so some useful comparisons can be made with systems such as GSM. It is accepted that 15 voice channels can be placed on each CDMA carrier, and as each carrier is 1.25 MHz wide (1.2288 MHz in exact terms), this represents a spectrum usage figure of 12 voice calls per megahertz. GSM can carry eight voice channels per carrier and there are five carriers per megahertz. This appears to provide a usage of 40 calls per megahertz. However, this is not the full story because neighbouring cells have to use different frequencies. With a cluster size of 12, i.e. 12 cells in a cluster, this figure reduces to 3.3 calls per megahertz, although when the frequency hopping facility of GSM is employed, cluster sizes of seven can be used and the frequency usage rises to six calls per megahertz. When comparing the two systems, other factors such as handover, which uses up some of the capacity of a CDMA system, are taken into account and it appears that cdmaOne has a capacity improvement of around 30 per cent. While much smaller than the original figures seemed to indicate, it is still very important when the level of calls that can be carried represents the earnings that can be made by the telephone company, and a figure of 30 per cent is very significant.

Another of the advantages of CDMA is that new cells can be added relatively easily without the need for a considerable amount of frequency planning. By adding new cells, further capacity is gained, albeit at the expense of additional interference because there are more mobiles and base stations that can interfere with one another. Although there is a loss of capacity in each cell, there are more cells and this brings an overall improvement in capacity, often by as much as 15 additional channels for the addition of a new base station. Fortunately by tailoring the antennas in each base station to provide the required coverage it is possible to limit the levels of overlap in the cells, and thereby reduce some of the causes of interference.

CDMA2000

CDMA2000 was devised in response to the effort to generate a global third generation mobile telecommunications system. While there are three main 3G standards that can be used, CDMA2000 has the advantage that it provides a full evolution from IS-95 (cdmaOne) through to the full high data rate 3G service. In this CDMA2000 the main evolution points are a progression from cdmaOne to the CDMA2000 1X, then 1xEV-DO (Evolution Data Only) and then 1xEV-DV (Evolution Data and Voice). The last two evolutions, i.e. EV-DO and EV-DV, are sometimes collectively referred to as CDMA2000 1xEV.

The various CDMA2000 systems will be described in the order of their evolution, showing the various additions and the path that their development follows.

The basic CDMA2000 system is available in two configurations, namely 1X and 3X. The 1X variant was first launched in South Korea in early 2001 and is now established there. While introducing many new facilities the new system is still backwards compatible with cdmaOne (IS-95). The 3X variant requires three times the bandwidth, i.e. 3.75 MHz.

For the forward link, the basic channels including the pilot, paging and sync channels are retained and provide backwards compatibility, but further channels are included to enhance the performance. These include a forward broadcast control channel that may be used for functions including paging. Although this facility was accessible via the paging channel on cdmaOne, the introduction of a new channel means that the congestion sometimes encountered on the paging channel is relieved. A forward auxiliary pilot channel is also included and this is used to assist in receiver decoding under poor conditions. A forward quick paging channel is another introduction. This is used to wake up a mobile in sleep or idle mode so that it can receive a message. The use of this channel enables the mobile to be kept in idle mode to conserve battery life.

New dedicated channels are also included. One named the forward dedicated control channel is introduced. This is used primarily for signalling and it can also be used for the retransmission of other high priority traffic including data. A second named the forward supplemental channel is used purely for data traffic and is able to handle much higher data rates than the forward fundamental channel used in cdmaOne (IS-95).

Modifications are also included for the reverse link. Two new common channels, namely the reverse common control channel, and the reverse enhanced access channel, are provided. These give more reliable access by allocating slots to users before they transmit bursts of data on the reverse link, and to provide a coherent form of reception for the base station probe receiver.

New dedicated channels are also included. These include a reverse pilot channel, a reverse supplemental channel and a reverse dedicated common control channel.

The next stage in the evolution of CDMA2000 is 1xEV. As both cdmaOne and CDMA2000 1X were developed with the primary intention of increasing voice capacity they are primarily optimized for circuit switched rather than packet switched data. The next evolution of CDMA2000 is intended to make the system more suitable for high speed data transmission and as a result multiple access packet switching is incorporated.

CDMA2000 1xEV-DO uses a hybrid CDMA/TDMA approach to provide high speed data only services. This requires a dedicated 1.25 MHz carrier on the forward link while the reverse link retains many of the features of the cdma2000 format. However, if voice services are to be provided as well an additional CDMA2000 carrier is required. The reason for adopting an evolution in this way is because it was envisaged that 1x EV-DO would be used in conjunction with data only applications with a PC where no voice link would be required. This means that if voice is required then two carriers are needed.

One of the main features of the 1xEV system is the inclusion of link adaptation. This means that users with a good signal at the base station will be able to have a higher data rate than a station further away for which the signal strength is much lower. Therefore stations are only able to use a higher data rate in situations where the system can succeed without a high rate of errors. By slowing the data rate down the errors are reduced. In addition to this the form of modulation is also adapted changing between QPSK, 8PSK and 16QAM, etc. to make the most efficient use of the bandwidth available. For the highest data rates 16QAM is used, although for this to operate with a low bit error rate a better signal to noise ratio is required than can be accepted by the other forms of modulation. Accordingly when the link quality falls, the data rate is slowed as other forms of modulation are used.

To achieve the level of link adaptation that is required, the reverse channels need to feedback a significant amount of information about the link quality. This requires additional channel capacity that is provided under the scheme.

For CDMA2000 1x EV-DV the additional capacity is required for the transmission of the voice channel. In 1x EV-DV this is included as part of the basic system, rather than needing an additional carrier for this. EV-DV is an evolution from 1X and does not use the TDMA elements that are included within the DO system. In fact 1xEV-DV is more similar to CDMA2000 1X than 1xEV-DO is. Unlike EV-DO, EV-DV is backwards compatible with 1X.

On the forward link, EV-DV has all the channels that 1X has, plus the Packet Data Channel (PDCH) and the Packet Data Control Channel (PDCCH). As the name suggests, the PDCH carries the high rate data. It is similar to, but not identical to, the data channel portion of the EV-DO system. Similarities include the encoding and interleaving functionality as well as the adaptable modulation types (QPSK, 8-PSK or 16-QAM) and the variable transmitted data rate (which is based on channel conditions). The PDCH configuration (data rate, modulation type, target mobile, etc.) can change every slot (1.25 ms). The PDCCH is used to broadcast the configuration of each PDCH slot.

Like the forward link, the EV-DV reverse link has all the channels of the 1X system. In addition, it has two new channels: the Acknowledgement Channel and the Channel Quality Indicator Channel. The mobile uses the Acknowledgement Channel to inform the base station whether the last packet has been received correctly or not. The CQICH is used to provide the base station with low latency information on the channel conditions. The base station uses the information from these channels (received 800 times per second) to determine which mobile should be serviced in the next slot and which packet should be transmitted to that mobile.

W-CDMA

The third generation system that is being adopted in Europe and many other places is known as W-CDMA (Wideband CDMA) or UMTS (Universal Mobile Telecommunications System). Although it uses a new method of modulation, much of the architecture is based on the very successful GSM system. However, it has been optimized to enable Internet data to be handled with ease. In common with other third generation systems, the emphasis is on the flexible delivery of high speed data as well as the normal voice connections and a general improved quality of service. It is also planned to use the exceedingly popular SMS concept. In this way messages will be able to be passed from one network system to another.

Obviously one of the main features of W-CDMA or UMTS is that it uses wideband CDMA to enable it to receive many different signals at the same time. GSM uses a mixture of frequency division multiple access (where different users are given different channels or frequencies) and time division multiple access (where users are allocated different timeslots). The W-CDMA system is set up so that the bandwidth of the signal is 5 MHz wide, but naturally this bandwidth can accommodate many different users. The bandwidth is much wider than that used with cdmaOne, for example, because a higher chip rate is used when encoding the CDMA signal, 3.84 MHz as opposed to 1.2288 MHz for cdmaOne.

While most systems allow either FDD or TDD to be used, within the UMTS specification both frequency division and time division duplex are allowed. Where frequency division duplex is used, a separation of 190

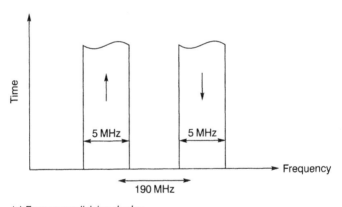

(a) Frequency division duplex

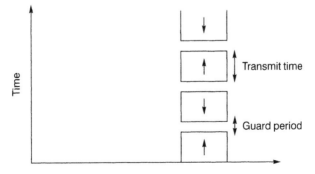

(b) Time division duplex

Figure 10.10 *Access methods for WCDMA*

MHz is used. This type of spectrum allocation is called paired spectrum as there is one band for the uplink and another for the downlink. In some areas of the world where spectrum allocations are limited time division duplex will be used. Here, bursts of data are transmitted in each direction; however, a guard period is required between transmitting. Naturally this type of spectrum allocation is known as unpaired spectrum. When using TDD, it is necessary to synchronize the base stations so that they transmit at the same time. This prevents a base station in an adjacent cell transmitting when a weaker mobile is transmitting. It is found that each type of system has its advantages. TDD is most suited to use with small cells where high data rates are needed, whereas FDD is better suited to covering wider areas.

The architecture required for UMTS is not the same as that required for GSM although work has been undertaken to ensure that the maximum reuse possible is available. In this way the investment required for the new infrastructure is reduced to a minimum. In the new system a new radio access network known as a UTRAN (UMTS Terrestrial Radio Access Network) is used. One of the major elements of the UTRAN is the base station or Node B as it is called. This communicates over a radio link with the mobile handsets and back into the network, as shown in Figure 10.11, over what is termed an I_{ub}-interface. The Node B provides all the

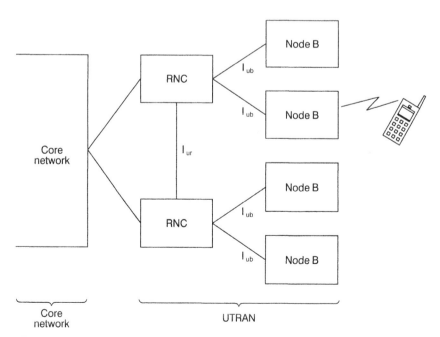

Figure 10.11 *The UMTS Terrestrial Radio Access Network*

RF processing to enable the transmission and reception of signals to and from the handsets.

The Node Bs are controlled by what is known as an RNC or Radio Network Controller. This is broadly equivalent to the base station controller found in GSM. It controls several Node Bs. It manages the resources, allocating the capacity for the data channels, enables facilities such as call set-up, and provides switching and traffic routing. Although it appears similar to a BSC it also includes some of the functionality of the MSC. This enables the handovers between cells to be handled solely by the UTRAN. In this way it is not necessary to involve the core network, even when cell handovers between different RNCs are required. This makes handovers much easier to manage. The RNC then communicates with the core network over an interface known as the I_u interface.

The location register databases that were found in GSM are broadly carried forward into UMTS. The Home Location Register (HLR) contains the subscriber information. It also contains information to enable changing and routing of messages to the current location of the mobile. It also allows for various location service related information to be accessed. There is the Authentication Centre (AuC) where data for each subscriber is stored. It essentially allows the user access to the network, checking elements such as the IMSI. Finally there is the Equipment ID Register (EIR). This database stores the Mobile Equipment Identities. The Mobile Equipment ID is the equivalent of the IMEI used in GSM. For simplicity these databases are often collectively called the Home Subscriber Server (HSS).

While many of the early mobile telecommunications services were located around 900 MHz with developments such as GSM 1800 and GSM 1900 taking advantage of higher frequency allocations, the spectrum reservations are all around the 2 GHz region. For Frequency Division Duplex (FDD) the uplink frequency reservation is between 1920 and 1980 MHz and for the downlink it is 2100 to 2170 MHz. This provides two 60 MHz wide reservations. Reservations are also in place for possible Time Division Duplex (TDD) operation. One is a 20 MHz wide slot between 1900 and 1920 MHz, and there is another optional 15 MHz wide band stretching between 2010 and 2025 MHz. As the FDD option is the one on which the first services will be launched, the text will focus on this mode of operation. However, it is not expected that all frequencies will be available in all countries.

As already mentioned the bandwidth allocation for each carrier or channel is 5 MHz. In line with other CDMA systems there are many mobile units transmitting within the same 5 MHz channel. The power of these is regulated by the base station or Node B so that the strength of each of them is approximately the same and in this way individual transmissions can be extracted as described in Chapter 3. All the mobile stations should be received by the base station with approximately the same signal strength for the system to be able to operate satisfactorily. Stations further away obviously will be received with a lower strength if no adjustment is made, whereas those that are close to the base station will be much stronger. If this is left uncorrected the weaker stations will be masked out by the stronger ones. This problem, often known as the 'near-far-problem' is resolved by transmitting power control information to the mobile equipment 1500 times a second. In this way the signal from each mobile can be regulated to be at the same level as the others.

In just the same way that cdmaOne and CDMA2000 were able to perform a soft handover, the same is true for a UMTS system. Close to a base station the mobile equipment communicates with only one base station. However, as it moves to the edge of a cell it is possible to communicate with two base stations because the channels are created by the use of different orthogonal codes when the CDMA signal is generated. No frequency change is required. It is therefore possible for communication to take place between the mobile and two base stations, and when communication is sufficiently reliable the mobile will only communicate with the second base station. If communication with the second base station worsens then the handover can be stopped.

A hard handover only occurs if the mobile has to change to a different protocol or frequency band. This may occur if the mobile reaches the edge of UMTS coverage and needs to revert to a GSM standard. This scenario will occur frequently during the early days of system roll-out when UMTS coverage is limited. Another scenario where a hard handover is required may occur if the mobile needs to change from an FDD to a TDD system. This may occur when moving from a region served by one operator to another, or if a TDD system is installed in a localized area to meet some specific requirements.

Cell breathing

One problem encountered with CDMA systems is known as cell breathing. When traffic in a cell increases, so does the interference level and as a result the power levels of the mobiles must increase. Particularly towards the edge of a cell where higher power levels are required anyway, a point is reached where the mobile cannot increase its power level any more and reception is lost. It can be seen that there will be areas where there is no coverage if traffic is high. To prevent this problem, careful network planning is required to ensure it can cope with the peak levels of traffic that will be encountered.

TD-SCDMA

Time Division Synchronous CDMA is a third generation cellular telecommunications system that has been developed in China. It uses unpaired channels, i.e. send and receive are on the same frequency and accordingly uses time division duplex to enable data to be sent in both directions. In order to enable the transmit–receive changeovers to occur at the same time so that interference between cells is reduced, all the cells are synchronized. Also the signals within the different base stations and mobiles are synchronized to ensure complete orthogonality of the spreading codes so that the different CDMA signals being transmitted at the same time can be effectively decoded with the minimum of interference between them. A further feature of the system that may be implemented in years to come is that smart antennas can be used. These will detect the approximate location of the mobile user and beam the signal in that direction. In this way power can be delivered only where it is needed and interference is reduced.

The system provides a variety of different data rates for different types of service. Voice data is at a rate of 8 kbps, while fax is transmitted at 9.6 kbps. For data applications a variety of data rates up to 2.048 Mbps can be used.

The basic system comprises the wireless base station (WBS) and multiple user terminals (UTs). For each basic WBS there is an RF channel that can provide up to 64 combined TDMA and CDMA code channels. These can be combined to provide a user data rate of up to 2 Mbps.

The RF carrier with sidebands occupies around 1.1 MHz at the − 3 dB point, and a 300 kHz guard band is allowed between neighbouring channels to allow adjacent channels to be used in the same cell. Two types of modulation are used. For data being transmitted at less than 384 kbps QPSK is used, but for rates above this it is modified for 16QAM.

Each of the RF channels has eight TDMA slots and then for each of these it is possible to have 16 CDMA code channels by providing the 16 CDMA orthogonal codes. The CDMA codes are generated from specific Walsh codes, and then a further pseudo-random spreading code is applied.

The system uses what is termed synchronous CDMA. This arises from the fact that to enable a very high level of interference reduction between mobiles, the spreading sequences are very accurately synchronized at the WBS. It is claimed that this virtually removes the co-channel interference from other code channels.

In line with other cellular systems a number of 'channels' are defined to carry the various types of information required. Four common channels are defined, namely the Broadcast Control Channel, Forward Access Channel, Paging Channel and the Random Access Channel. The dedicated channels include the Dedicated Control Channel and the Dedicated Traffic Channel.

Looking at the channels individually, the Broadcast Control Channel is a downlink channel that is used to broadcast system and cell specific information as well as messages such as time, weather, public information, etc., to the users in the cell.

The Forward Access Channel is a downlink channel used to carry control information to a UT after the uplink access control signal is received. The channel may also carry short packets of information for the user. The Paging Channel, as the name suggests is used to carry information to a UT that is in standby. The final control channel is an uplink channel. Named the Random Access Channel it is used to carry control information from a user to the WBS. It may also carry other short data packets.

As already mentioned there are two dedicated channels. These are present in both uplink and downlink. The first is the Dedicated Control Channel. This is used to carry a variety of types of control information to and from the UT to enable it to operate satisfactorily within the cell.

The other dedicated channel is the Dedicated Traffic Channel. In many respects this is the most important one as it carries the actual user data in both directions between the WBS and the UT.

One of the future features of TD-SCDMA is the use of SMART antennas. These antennas will use the information gained about the location of the user to direct the signal for a particular user in the required direction. In this way levels of interference will be reduced, thereby ensuring that data rates can be maintained and capacity held at its maximum. However, this feature is not being incorporated in the early developments of TD-SCDMA.

Fourth generation

In view of the long period that it takes between the first ideas and the final roll-out of systems, the initial concepts for the fourth generation systems are being set in place. These will take many of the lessons learned from the roll-out of the third generation systems, and obviously considerably enhance the performance and add new features that are considered likely to be required. In addition to this they will take account of the developments in technology that will occur before their roll-out.

11 Short-range wireless data communications

In recent years a number of new technologies have arisen that provide short-range data communications for such applications as providing a wireless local area network, or just some connectivity between a computer and a printer without the need for a wired connection. Such technologies as Bluetooth, HomeRF and IEEE 802.11 are the main contenders.

Each of these has its own advantages and is aimed at a slightly different market. Bluetooth technology is aimed at providing short range, often ad hoc connections between units. It is designed so that it can be designed into an application specific IC (ASIC) to reduce its cost and also keep its size down. It is also aimed at the cellular telecommunications market where it is expected to be included in cellular phones so that they can easily connect to items such as laptop computers. The HomeRF technology, as the name indicates, is aimed very much towards applications in the home. It will be able to provide connectivity for multiple computers in the home onto a single Internet access point, generally computer interconnectivity, and many other applications. Its main features are its simplicity, low cost and its scalability, enabling voice also to be carried over it. Finally there is IEEE specification 802.11. This is aimed very much at the business market where performance is the key factor. It is being incorporated into many laptop computers to provide features such as roaming, while also giving a high degree of security.

The different technologies share many things in common, although they are not able to interface with one another. One of the major common aspects is that they all work in the industrial, scientific, and medical portions of the spectrum for which no licence is required. The main band that is used is the 2.4 GHz band. The others are 915 MHz and 5 GHz. As these bands do not require a licence, and they have many users, even microwave ovens use frequencies in the band. Accordingly, all three systems have methods of being able to avoid the high levels of interference that are to be found there.

Bluetooth

This technology traces its origins back to 1998 when a special interest group was formed by IBM, Intel, Nokia and Toshiba. The name of the standard originates from the Danish king Harald Blatand who was reputed to have united the Scandinavian people in the tenth century AD. The Bluetooth standard took its name from him because it endeavours to unite personal computing devices.

Bluetooth is primarily a data system and can carry data at speeds up to 721 kbps, but it also offers up to three voice channels. The technology enables a user to replace cables between devices such as printers, fax machines, desktop computers and peripherals and a host of other digital devices. Furthermore, it can provide a connection between an ad hoc wireless network and existing wired data networks.

The system is based round a frequency hopping carrier. This provides a greater dynamic range than if a direct sequence spread spectrum technique had been adopted. If direct sequence spread spectrum techniques were used then other close-by transmitters could block the required transmission if it is further away and weaker. The data is modulated onto the carrier using a form of frequency shift keying known as Gaussian frequency shift keying. The frequency shift keyed carrier is filtered to ensure the sidebands do not extend too far either side of the main carrier. By doing this it achieves a bandwidth of 1 MHz with stringent filter requirements to prevent interference on other channels.

With a hopping transmission, the carrier only remains on a given frequency for a short time, and if any interference is present the data will be resent later, but on a different channel that is likely to be clear of other interfering signals. The standard uses a hopping rate of 1600 hops per second. These hops are spread over 79 fixed frequencies that are chosen in a pseudo-random sequence. The fixed frequencies occur at $2400 + n$ MHz where the value of n varies from 1 to 79. This gives frequencies of 2402, 2404 . . . 2480 MHz. In some countries the ISM band allocation does not allow the full range of frequencies to be used. In France, Japan and Spain, the hop sequence has to be restricted to only 23 frequencies.

The transmitter powers for Bluetooth are usually low. Three classes are defined in the specification, 1 mW, 2.5 mW and 100 mW. Most are either the first two classes that provide ranges of up to 100 mm and 10 m respectively.

In order to enable effective communications to take place in an environment where a number of devices may receive the signal, each device has its own identifier. This is provided by having a 48-bit hard wired address identity giving a total of 2.815×10^{14} unique identifiers and this should be quite sufficient!

Using Bluetooth there are two types of link that are possible. There is the Asynchronous Connectionless Communications Link (ACL) and this is used for file and data transfers. A second form of link is known as a Synchronous Connection-orientated Communications Link (SCL) and two of these are available. These are used for applications such as digital audio. The asynchronous link supports a maximum data rate of 732.2 kbps in an asymmetric mode, whereas in a symmetrical mode running the same data rate in both directions this rate is reduced to 433.9 kbps. The synchronous links support two bi-directional connections at a rate of 64 kbps. The data rates are adequate for audio and most file transfers. However, the available data rate is insufficient for applications such as high rate DVDs that require 9.8 Mbps or for many other video applications including games spectacles.

Data is organized into packets to be sent across the link. The Bluetooth specification lists 17 different formats that can be used dependent upon the requirements. They have options for elements such as forward error correction data and the like. However, the standard packet consists of a 72-bit access code field, a 54-bit header field, and then the data to be transmitted which may be between 0 and 2745 bits. This data includes the 16-bit CRC if it is needed.

As it is likely that interference will cause errors, error handling is incorporated within the system. For asynchronous links packet sequence numbers are transmitted. If an error is detected in a packet then the receiver can request it to be resent. Error coding using a 16-bit CRC is also available. For the synchronous links packets cannot be resent as there is unlikely to be sufficient bandwidth available to resend data and 'catch up', although it is possible to include some forward error control.

When communicating Bluetooth devices form small nets called 'piconets'. These comprise up to eight devices, of which one takes on the role of a master while all the others become slaves. If more than eight devices are within range, they may remain in an inactive standby state and may be requested at a later time to join the net. Still further devices may be in a standby state.

When establishing a net, the master transmits an enquiry message every 1.28 seconds to discover whether there are any other devices within range. If replies are received then an invitation to join the net is transmitted to specific devices that might be in range. To set up the net the master allocates each device a member address and it then controls their transmissions.

All Bluetooth devices have a clock that runs at twice the hopping speed and this provides synchronization to the whole net. The master transmits in the even numbered timeslots while the slaves transmit in the odd numbered slots once they have been given permission to transmit.

It is possible for Bluetooth devices to encrypt data. A key up to 128 bits is used and it is claimed that the level of security provided is sufficient for financial transactions. However, in some countries the length of the key is limited to enable the security agencies to gain access if required.

IEEE 802.11

There is a family of '802' specifications relating to local area networks. For example, 802.3 is the Ethernet specification, although 802.11 is reserved for wireless local area networks (WLANs) and this is further subdivided. Of these 802.11b is becoming one of the most popular for computer and other similar applications. It has been developed to enable high performance radio links to be established to support roaming in large offices or other similar environments and to reduce the wire inter-connections that are required in any office environment.

The performance of 802.11b is high, supporting data rates of up to 11 Mbps, and to achieve this it uses direct sequence spread spectrum techniques with 52 carriers occupying a total bandwidth of 26 MHz. Like Bluetooth it operates in the 2.4 GHz ISM band. Another specification designated 802.11a uses the 5 GHz band, supports data rates of up to 54 Mbps and employs orthogonal frequency division multiplex (OFDM). It is gaining some popularity because it is believed that the 2.4 GHz band is likely to become very congested. However, as 802.11b is currently by far the most popular, this will be described here.

Using the 802.11 system there are two types of network that can be formed, namely what are termed infrastructure networks and ad hoc networks. The infrastructure systems have a wire backbone that is connected to a server and uses the wireless link to enable connection to the full computer network and other local services such as printers. The wireless network is then split up into a number of cells, each serviced by a base station or access point (AP) which provides access to the backbone of the full system. This access point may have a range of between 30 and 300 metres dependent upon the environment.

The wireless system will enable considerable savings to be made in not having to install costly and unsightly wiring to each computer in a network, and it also provides the flexibility to physically move desks in an office without the need to move all the wiring. It also provides almost instant access for people bringing laptop computers into an office to connect without having to make physical wire connections each time.

Ad hoc networks may also be used. Here computers and peripherals may be brought together and communicate with each other without being part of a larger system. This type of configuration does not need the use of an access point. For this type of system, algorithms have been

Wired backbone distribution system

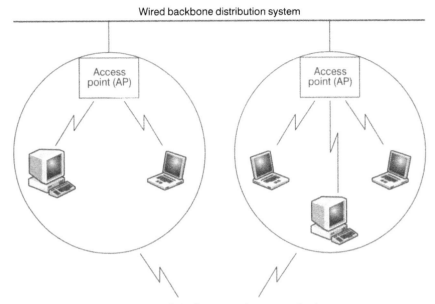

Approximate limit of access point communications

Figure 11.1   *IEEE 802.11 system as used to provide connectivity for a large network system*

written to enable one of the computers to elect a base station or master, with the others remaining as slaves. This type of network may be established when several people come together and need to share data between their computers or if they need to access a printer without the need to use wire connections.

Unlike Bluetooth, 802.11b uses direct sequence spread spectrum. This allows it to operate in an environment with other users, provided that the levels of interference do not rise too high. A variety of different types of modulation are used. Systems running at 1 Mbps use BPSK (binary phase shift keying) modulation, and those running at 2 Mbps QPSK (quaternary phase shift keying) modulation. For higher data rates, systems running at 5.5 Mbps and 11 Mbps use CCK (complementary code keying) and QPSK modulation. CCK involves 64 unique code sequences, each of which supports 6 bits per codeword. The CCK codeword is then modulated onto the RF carrier using QPSK, and this allows another 2 bits to be encoded for each 6-bit symbol. In this way each 6-bit symbol contains 8 bits, i.e. 1 byte.

Around the world different countries have slightly different legislation and frequency allocations. As a result there are differences between the channels and hence centre frequencies that are used. A summary is given

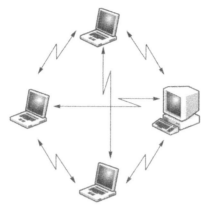

Figure 11.2 *An example of an ad-hoc network*

in Table 11.1. FCC channel frequencies are used within USA and the same frequencies are available for use in Canada. ETSI frequencies are adopted within Europe, although frequencies in France are limited to channels 10–13. To prevent interference between neighbouring access points there should be a 30 MHz separation between them although there is a move to use a 25 MHz separation.

The 802.11 system is symmetrical with all stations having the same power output capability. This must be the case because the system must be able to operate in an ad hoc mode. Like the different frequency

Table 11.1 *Channels frequencies and identification*

Channel ID	FCC Channel Frequencies	ETSI Channel Frequencies
1	2412 MHz	N/A
2	2417 MHz	N/A
3	2422 MHz	2422 MHz
4	2427 MHz	2427 MHz
5	2432 MHz	2432 MHz
6	2437 MHz	2437 MHz
7	2442 MHz	2442 MHz
8	2447 MHz	2447 MHz
9	2452 MHz	2452 MHz
10	2457 MHz	2457 MHz
11	2462 MHz	2462 MHz
12	N/A	2467 MHz
13	N/A	2472 MHz

allocations, different countries have different power limitations. For spread spectrum modes the maximum transmission power is generally specified to be 100 mW (EIRP). However, in the USA this can rise to 1000 mW and in Japan the power is specified slightly differently as 10 mW/MHz. In any country the minimum power level does not fall below 1 mW. Below this level reliable communications would not usually be possible. Within these limits different power levels are set by the system so that the minimum power level is chosen consistent with reliable communications. In this way interference to nearby cells is minimized, and it also conserves battery power for items like laptop computers. Also the system will reduce the data rate if errors start to be detected. Even so it is generally accepted that the maximum reasonable distance over which 802.11b will operate is 100 metres for a clear line of sight path.

In terms of the actual operation of the system, when a station wants to join a cell it must synchronize with the other stations in the cell and it must be allowed in. The synchronization can be gained by two means. The first is known as passive scanning where the station waits to receive what is termed a beacon frame from the access point or another station in an ad hoc network. This beacon frame is a periodic frame of data that is sent out with synchronization information on it. Apart from enabling new stations to synchronize, they also ensure that stations already in the cell maintain their synchronization. When entering a cell the station may alternatively send out a probe transmission and then wait for a response.

Once the new station has the synchronization information it can then request access to the cell. It does this by going through an authentication process. Only when this has been completed can the station enter the cell and exchange data. The process includes exchanging passwords to ensure that the new station should be in the cell.

When transmitting, the system uses a collision avoidance technique known as CSMA/CS and this is combined with a positive acknowledge. When a station wants to transmit, it checks the frequency to ensure no other stations are transmitting and if the channel is occupied the station delays sending its message for a random amount of time. When all is clear it sends its message. As there is a possibility another station could perform the same operation at the same time, an acknowledgement is sent to the station from whom the message is received. If the station sending the message does not receive its acknowledgement then it resends its message.

The first message that is sent is what is termed a request to send (RTS). This is done in case there are other stations that cannot be heard. This RTS message defines the length of the message to be sent and the destination. If this is acceptable a clear to send (CTS) message is sent.

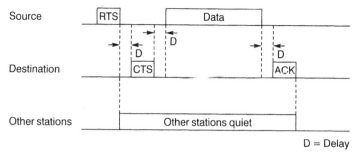

Figure 11.3 *Transactions between stations transmitting data to each other*

In fact anything sent across the system is split into frames. There are three types of frame. Data frames that are used to send actual data from one station to another, control frames that are used to control the way in which data is sent (examples of these are the RTS, CTS and ACK frames that are sent) and finally there are the management frames. These are sent across the system in the same way as the data frames and they are used to control or manage the cell, but the data is not visible to the user.

HomeRF SWAP

The HomeRF Shared Wireless Access Protocol (SWAP) is designed primarily for home applications where it is intended to carry both voice and data information for a broad range of interoperable consumer devices. The system is intended for a broad range of items including PCs, peripherals, cordless phones and many other devices to share and communicate voice around the home without the need to run fixed and expensive wiring. Unlike Bluetooth, which is aimed primarily for use with cellular technology, or IEEE802.11 which is intended for business use, SWAP is intended for home use where the developers see a vast and growing market.

The system can support data rates of between 10 and 20 Mbps combined with sufficient range for most residential applications. If operating over slightly greater distances the speed can back off to 5 Mbps. Like 802.11b and Bluetooth, HomeRF operates in the 2.4 GHz ISM band. It uses frequency hopping with a hop rate of between 50 and 100 hops per second.

The system can be set up in either ad hoc, peer to peer mode, or as a network with a control point (CP) giving access to a wired network within the house. In this respect it has many similarities to 802.11b.

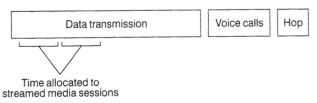

Figure 11.4   *HomeRF data frame structure*

When requiring to send data the CSMA/CS protocol derived from 802.11 is used. Here the unit wanting to send data listens to see if the frequencies are clear to send data. If it is clear then the data is sent, otherwise it waits for a random amount of time before trying to send the data again.

The data within a HomeRF transmission is contained within repeating frames that are transmitted. Each one of these frames is either 10 or 20 ms long dependent upon the number of active voice calls at any given time. Within the frame, the bulk of the available capacity is used for asynchronous data transmission. However, within the capacity reserved for this, priority is given to the streaming media sessions. Up to eight of these sessions are allowed, but when fewer than this number are present, the available capacity remaining can be used for ordinary data transmission.

The last part of the data frame structure is reserved for providing voice quality communications. These full duplex pairs of phone quality slots use the Digital Enhanced Cordless Telecommunications (DECT) system. The HomeRF protocols map directly onto this allowing an already well-established system to be incorporated into the HomeRF technology. The full specification for HomeRF allows for a total of eight land line quality links to be supported.

Index

802.11, 282, 285–289

Access point, 285
Active antenna, 114
Ad hoc network, 285
Advanced mobile phone system, 248
AFC, 155
AFSK, 62
AGC, 133–135
AM, 53–56
AM broadcasting, 204–207
AM demodulator, 152
AM transmitter, 195
Ampere, 4
Amplifier, 182–188
    class A, 183
    class B, 183
    class C, 184
    efficiency, 182
    IF, 148
    RF, 144
Amplitude modulation, 53–56
    modulation detector, 53
    modulation transmitter, 195
AMPS, 248
Analogue cell system, 262
Analogue to digital converter, 162
Angle of radiation, 37, 83
Antenna:
    active, 114
    bandwidth, 76, 82
    beamwidth, 81
    dipole, 95–98
    discone, 104
    end fed wire, 99
    ferrite rod, 112
    folded dipole, 97

gain, 79
impedance, 78
location, 115
log periodic, 105–107
loop, 113
materials, 116
parabolic reflector, 110–112
phase array, 99–101
polarization, 78
resonance, 76
safety, 118
short, 109
vertical, 107–109
Yagi, 101–104
Apogee, 223
Armstrong, Edwin, 10, 11, 13
ASCII, 63
Atmosphere, 30–33
Audio filter, 165
Audio frequency shift keying, 62
Audion, 9
Aurora, 43
Automatic frequency control, 155
Automatic gain control, 133–135
Automatic volume control, 133
AVC, 133

Balanced feeder, 92
Bandwidth (of FM signal), 61
Bandwidth antenna, 76, 82
Bandwidth filter, 166–168
Base station, 258
    controller, 261, 264
    subsystem, 264
    transceiver sub-system, 264
BBC, 12
Beamwidth antenna, 81

Beat frequency oscillator, 53, 152
Bell, Alexander Graham, 5
BFO, 53, 152
Blackout radio, 43
Blocking, 177
Bluetooth, 282, 283–285
Broadcast bands short wave, 206
Broadcast bands tropical, 206
Broadcasting AM, 204–207
Broadcasting digital, 215–220
Broadcasting RDS, 211–215
Broadcasting short wave bands,
    204–207
Broadcasting VHF FM, 207
BSC, 261, 264
BSS, 264
BTS, 264
Butterworth filter, 190

Capture effect, 156
Carrier, 51
Carrier insertion oscillator, 152
Cat's whisker, 10
CDMA, 15, 72–75, 249
CDMA2000, 249, 273–275
cdmaOne, 15, 249, 269–273
Cell, 254
Cell breathing, 279
Ceramic filter, 150, 165
Chip code, 73
CIO, 152
Clarke, Arthur C., 14, 221
Class A amplifier, 183
Class B amplifier, 183
Class C amplifier, 184
Coaxial feeder, 90
Coded orthogonal frequency
    division multiplex, 75
COFDM, 75
Coincidence demodulator, 159–161
Communications satellite, 231–235
Continuous tone coded squelch
    system, 239
Cooke, 4
Critical frequency, 40
Cross modulation, 177
Crystal filter, 149, 165

Crystal mixer oscillator, 180
Crystal set, 120
CTCSS, 239

D layer, 31, 32
DAB, 215–220
DBS, 234
DDS, 141–144
De Forest, 9
De Magnete, 3
Dead zone, 41
De-emphasis, 61, 207
Demodulation:
    FM, 153–162
    synchronous AM, 152
Demodulator, 151–162
    AM, 53, 152
    coincidence, 159–161
    Foster-Seeley, 158
    Morse, 152
    phase locked loop FM, 161
    quadrature, 159
    ratio, 157
    slope, 156
    SSB, 152
Detector, AM, 53
Deviation, 58
Deviation ratio, 60
Diffraction (of radio waves), 26–28
Digital broadcasting, 215–220
Digital signal processing, 162
Digital to analogue converter, 143,
    162
Diode detector, 53
Diode ring mixer, 144
Diode valve, 8
Dipole antenna, 95–98
Dipole, folded, 97
Direct broadcast satellite, 234
Direct conversion receiver, 123–127
Direct digital synthesizer, 141–144
Direct sequence spread spectrum,
    72–75
Director, 102
Discone antenna, 104
Discontinuous transmission, 267
Divider frequency, 138

DMO, 244
DSP, 162
DSSS, 72–75
DTMF, 239
DTX, 267
Dual tone multiple frequency, 239
Dynamic range, 177

E layer, 31, 33
EDGE, 249, 268
Edison, 5, 8
Efficiency of amplifier, 182
Electric field, 17–19
Electromagnetic wave, 20, 21
Electromagnetic wave spectrum, 22
Elliptical orbit, 223
End fed wire antenna, 99
Enhanced date rates for global
    evolution, 249, 268
Enhanced other networks, 214
EON, 214
EV-DO, 273–275
EV-DV, 273–275

F layer, 31, 33, 34
Fading, 42
Faraday, 4
Faraday rotation, 231
FDD, 256, 276
FDMA, 249
Feeder, 84–94
    balanced, 92
    coaxial, 90
    impedance, 85
    loss, 89
    twin, 92
Ferrite rod antenna, 112
Filter, 188–191
    audio, 165
    bandwidth, 166–168
    Butterworth, 190
    ceramic, 150, 165
    crystal, 149, 165
    IF, 148–151
    LC, 149, 165
    pass-band, 167

shape factor, 168
stop-band, 168
Fleming, John Ambrose, 7, 8, 9
FM, 58–60
    broadcasting, 207
    demodulation, 153–162
    transmitter, 199
Folded dipole, 97
Foster-Seeley demodulator, 158
Frequency, 21
Frequency divider, 138
Frequency division duplex, 256, 276
Frequency division multiple access,
    249
Frequency hopping, 71
Frequency modulation, 58
Frequency modulation transmitter,
    199
Frequency multiplier, 185
Frequency shift keying, 62
Frequency synthesizer, 135–144, 181
FSK, 62

Gain antenna, 79
Gaussian filtered minimum shift
    keying, 65–67, 266
General packet radio system, 249,
    268
Geocentre, 223
Geostationary orbit, 223, 225
Gilbert, William, 3
Global system for mobile
    telecommunications, 248,
    263–268
GMSK, 65–67, 266
GPRS, 249, 268
GPS, 235
Ground wave, 34
Groundtrack, 223
Groupe speciale mobile, 248
GSM, 15, 248, 263–268

Harmonics, 190
Henry, 4, 5
Hertz, Heinrich, 5, 21
HLR, 265, 278
Hohmann transfer principle, 228

Home location register, 265, 278
Home subscriber server, 278
HomeRF, 282, 289–290

IF amplifier, 148
IF breakthrough, 163
IF filter, 148–151
Image response, 128, 131
IMEI, 264
Impedance antenna, 78
Impedance feeder, 85
Impedance output, 193
IMSI, 264
Intermodulation, 191
Intermodulation distortion, 175
International mobile equipment
    identity, 264
International mobile subscriber
    identity, 264
Ionization, 31
Ionosonde, 41
Ionosphere, 31–34
Ionospheric disturbances, 43
Ionospheric propagation, 35
I-Q modulator, 66
IS-95, 269–273
Isotropic source, 80

Kemp, 7

Latour, 10
LC filter, 165
Levy, Lucien, 11
Local oscillator, 146
Lodge, Sir Oliver, 6
Log periodic antenna, 105–107
Loop antenna, 113
Lowest usable frequency, 41
LUF, 41

Macrocell, 255
Magnetic field, 19, 20
Marconi, 6, 7
Master oscillator, 179

Matching network, 188–191
Maximum usable frequency, 40
Maxwell, 5
MDS, 177
Meteor burst communication, 48
Meteor scatter, 48
Meteor showers, 49
Meteors, 48
Microcell, 255
Minimum discernible signal, 177
Minimum shift keying, 64
Mixer, 144–146, 181
Mixer diode ring, 144
Mixing process, 123
Mobile equipment ID, 278
Mobile phone, 257
Mobile station, 264
Mobile switching centre, 261, 264
Mobile telephone exchange, 261
Modem, 62
Modulation, 51
    amplitude, 53–56
    frequency, 58
    index (AM), 57
    index (FM), 60
    phase, 63, 64
    pulse, 67–71
    pulse amplitude, 68
    pulse code, 69
    pulse width, 69
Modulator I-Q, 66
Modulator phase, 202
Morse, 52, 53
Morse demodulator, 152
Morse, Samuel, 4
Morse telegraph, 4
Morse transmitter, 194
MPT 1327, 240
MS, 264
MSC, 261
MSK, 64
MTX, 261, 264
MUF, 40
Multiple reflections, 38–40
Multiplier frequency, 185
Murray code, 63
Narrow-band FM, 59
NBFM, 59

Newton, Isaac, 221
NMT, 248
Node B, 277
Noise, 169
Noise factor, 173
Noise figure, 173
Nordic mobile telephone, 248

Oersted, 4
OFDM, 75
Ohm, 4
Orbit satellite, 222–225
Orthogonal code, 72
Orthogonal frequency division
    multiplex, 75
Oscillator, 179
Oscillator crystal mixer, 180
Oscillator variable frequency, 180
Output impedance, 193

Paget, 7
PAM, 68
Parabolic reflector, 110–112
Pass-band, 167
Path calculation, 229
Path loss, 26
PCM, 69
PDO, 244
Perigee, 223
Phase, 136
Phase accumulator, 142
Phase comparator, 137
Phase locked loop, 136
Phase locked loop FM demodulator,
    161
Phase modulation, 63, 64
Phase modulator, 202
Phase noise, 164
Phase shift keying, 64
Phased array antenna, 99–101
Phasing method of SSB generation,
    198
Picocell, 255
PLL, 136
PM, 63, 64
Polar diagram, 79, 82

Polarization, 24, 25
Polarization, antenna, 78
Power, 186, 187
Pre-emphasis, 61, 207
Programmable divider, 138
PSK, 64
Pulse amplitude modulation, 68
Pulse code modulation, 69
Pulse modulation, 67–71
Pulse width modulation, 69
PWM, 69

Quadrature demodulator, 159
Quieting, 155

Radiation angle, 83
Radio data service, 211–215
Radio network controller, 278
Radio spectrum, 22, 23
Radio wave, 20
Ratio demodulator, 157
Receiver:
    automatic gain control, 133
    crystal set, 120
    direct conversion, 123–127
    multiple conversion, 132
    sensitivity, 169–174
    superhet, 127–133
    TRF, 121
Reciprocal mixing, 164
Reflected signals, 29
Reflection (of radio waves), 26, 27
Reflector, 102
Refraction (of radio waves), 26–28
Regenerative detector, 122
Resonance, antenna, 76
RF amplifier, 144
RNC, 278
Round, H.J., 10, 11

Satellite communications, 231–235
Satellite, direct broadcast, 234
Satellite, navigational, 235
Satellite orbit, 221–225

Satellite weather, 236
Satellites, 225–228
SCA, 212
Selectivity, 166–169
Sensitivity, 169–174
SFN, 219
Shape factor, 168
Shared wireless access protocol, 289
Short antenna, 109
Short message service, 264, 267
Short wave broadcast bands, 206
Short wave broadcasting, 204–207
SID, 43
Sideband, 55
Sidebands (of FM signal), 60
Signal to noise ratio, 171
Signal to noise ratio improvement, 61
SIM, 264
SINAD, 172
Single frequency network, 219
Single sideband, 57, 58
Single sideband transmitter, 196
Skip zone, 41
Skywaves, 35
Slope demodulator, 156
SMS, 264, 267
Soft handover, 272
Sporadic E, 47
Spread spectrum, 71
Spurious outputs, 193
Spurious signals, 164
Sputnik, 14, 221
Squelch, 156
SSB, 57, 58
    demodulator, 152
    transmitter, 196
Standing wave ratio, 88
Standing waves, 85–88
Stereo broadcasts, 207–211
Stop-band, 168
Subscriber identity module, 264
Subsidiary communications
    authorization, 212
Sudden ionospheric disturbance, 43, 44
Sunspot cycle, 33
Sunspots, 32, 33

Superhet receiver, 127–133
SWAP, 289
SWR, 88, 187
Synchronous AM demodulation, 153
Synthesizer, 135–144, 181
Synthesizer, direct digital, 141–144
Synthesizer, multi-loop, 140

TACS, 248
TDD, 256, 276
TDMA, 249
TD-SCDMA, 249, 280
Telstar, 14
TETRA, 239, 243–246
Third order intercept, 176
Time division duplex, 256, 276
Time division multiple access, 249
Total access communications system, 248
Transmitter:
    AM, 195
    FM, 199
    SSB, 196
Transponder, 232
TRF receiver, 121
Tropical bands, 206
Troposcatter, 46
Troposphere, 31
Tropospheric ducting, 46
Trosopheric propagation, 45
Trunking, 240
Tuned radio frequency receiver, 121
Twin feeder, 92

Umbrella cell, 255
UMTS, 249, 273–279
    terrestrial radio access network, 277
Universal mobile
    telecommunications system, 249
UTRAN, 277

Variable frequency oscillator, 147, 180
VCO, 137

Velocity factor, 89
Vertical antenna, 107–109
Very low frequency propagation, 44
VFO, 147, 180
VHF broadcasting, 207
VHF propagation, 45
Visitor location register, 266
VLR, 266
Voltage controlled oscillator, 137
Vyvyan, 7

Waveform map, 142
Waveguide, 93
Wavelength, 21
WBFM, 59
W-CDMA, 275–279
Weather satellite, 236
Wheatstone, 1
Wide-band FM, 59

Yagi, 101–104

Printed and bound by CPI Group (UK) Ltd, Croydon, CR0 4YY

03/10/2024

01040432-0004